Approach your exams the IB way

Physics HL

David Homer

International Baccalaureate
Baccalauréat International
Bachillerato Internacional

IB Prepared

Approach your exams the IB way

Physics HL

Published February 2011

International Baccalaureate
Peterson House, Malthouse Avenue, Cardiff Gate
Cardiff, Wales GB CF23 8GL
United Kingdom
Phone: +44 29 2054 7777
Fax: +44 29 2054 7778
Website: http://www.ibo.org

The International Baccalaureate (IB) offers three high quality and challenging educational programmes for a worldwide community of schools, aiming to create a better, more peaceful world.
© International Baccalaureate Organization 2011

IB merchandise and publications can be purchased through the IB store at http://store.ibo.org. General ordering queries should be directed to the sales and marketing department in Cardiff.

Phone: +44 29 2054 7746
Fax: +44 29 2054 7779
Email: sales@ibo.org

British Library Cataloguing in Publication Data.
A catalogue record for this book is available from the British Library.

ISBN: 978-1-906345-29-7

Cover design by Pentacor**big**
Typeset by Wearset Ltd
Printed and bound in Spain by Edelvives

Item code 4056

2015 2014 2013 2012 2011
10 9 8 7 6 5 4 3 2 1

Acknowledgments
Diane Lloyd and Andreas Tsokos for their advice on the IB Diploma Programme physics higher level.

IB learner profile

The aim of all IB programmes is to develop internationally minded people who, recognizing their common humanity and shared guardianship of the planet, help to create a better and more peaceful world.

IB learners strive to be:

Inquirers   They develop their natural curiosity. They acquire the skills necessary to conduct inquiry and research and show independence in learning. They actively enjoy learning and this love of learning will be sustained throughout their lives.

Knowledgeable   They explore concepts, ideas and issues that have local and global significance. In so doing, they acquire in-depth knowledge and develop understanding across a broad and balanced range of disciplines.

Thinkers   They exercise initiative in applying thinking skills critically and creatively to recognize and approach complex problems, and make reasoned, ethical decisions.

Communicators   They understand and express ideas and information confidently and creatively in more than one language and in a variety of modes of communication. They work effectively and willingly in collaboration with others.

Principled   They act with integrity and honesty, with a strong sense of fairness, justice and respect for the dignity of the individual, groups and communities. They take responsibility for their own actions and the consequences that accompany them.

Open-minded   They understand and appreciate their own cultures and personal histories, and are open to the perspectives, values and traditions of other individuals and communities. They are accustomed to seeking and evaluating a range of points of view, and are willing to grow from the experience.

Caring   They show empathy, compassion and respect towards the needs and feelings of others. They have a personal commitment to service, and act to make a positive difference to the lives of others and to the environment.

Risk-takers   They approach unfamiliar situations and uncertainty with courage and forethought, and have the independence of spirit to explore new roles, ideas and strategies. They are brave and articulate in defending their beliefs.

Balanced   They understand the importance of intellectual, physical and emotional balance to achieve personal well-being for themselves and others.

Reflective   They give thoughtful consideration to their own learning and experience. They are able to assess and understand their strengths and limitations in order to support their learning and personal development.

Table of contents

1. Introduction

As an IB student you have many resources available on the path to your final exams in physics. This book is just one of the resources that will help you to prepare for the exam papers.

How to use this book

This book is one of the resources that you should use as you work towards the IB physics higher level exam. It cannot replace your notes or your textbooks; it certainly cannot replace your teacher. You must ensure that you learn all the material for the exams, as not every topic is covered in this book.

The book contains examples of answers that were written by IB students and which are accompanied by comments written by a senior examiner. There are also examples taken from recent multiple-choice papers with hints on how to answer them. You may wish to read the notes on "How do I approach the question?" and then try to answer the questions before reading the answers written by other students.

What is in this book?

- **Chapter 2** tells you what to expect in the three exam papers. The **structure** of each paper is explained, and you are told what the examiners look for in each of the papers.

- **Chapter 3** lists the **command terms** used in IB physics exams. The meaning of the most common terms is explained, and you are told what is expected of answers when the terms are used in the exam.

- **Chapter 4** contains **top tips**. These are pieces of advice specific to physics exams that will help you to gain marks.

- **Chapter 5** provides you with **key advice** for studying and revising physics. It gives strategies for revision, and gives you advice on using a published IB markscheme.

- **Chapters 6–13** match the **syllabus topics** as printed in the Diploma Programme *Physics guide*, with additional higher level material (AHL) included.

- Each of **chapters 6–13** begins with a list of the **key terms** that you should know or be able to define. There is sometimes an explanation of the term, but you should memorize the definitions of all these terms before the exams.

- Each chapter continues with sections devoted to the separate topics. You are provided with two lists of key ideas: "**You should know**" and "**You should be able to**". Part of your revision strategy should be to master these concepts. The "**Be prepared**" section comments on ideas that sometimes give difficulty to students. Additionally, there may be a question taken from a past paper 1 to illustrate how the topic can be examined in a multiple-choice test.

- Finally, each chapter contains a number of past paper 2 questions. For each of these extended questions, you will see the question itself first, followed immediately by

help on how you might attempt the question—these are simply suggestions, as there is often more than one way to gain credit in a question. You might choose to answer the question yourself before looking at the examples of work written by other IB students. There are three examples of each question at different levels, low, medium and high. For each of the examples, there is a commentary explaining why the student gained (⊕) or lost (⊖) marks, helping you to avoid pitfalls and offering you tips to improve your own marks (✔).

- **Chapters 14–19** contain the option topics, in roughly the same format as earlier chapters.

- **Chapter 20** contains other exam questions, so that you can go on to test yourself and your revision, putting into practice what you have learned from this book.

2. Get to know your exam papers

This chapter introduces you to the main features of the exam papers in physics. The structure of each paper is explained, and there are suggestions about tackling each paper.

General

- There are three papers in physics.

- The three make up 76% of the total mark; the remainder comes from your internal assessment work.

- For each paper, you will receive a copy of the *Physics data booklet*. There is more advice on using this booklet in chapter 4.

- It is important that you try to complete as much of each paper as you can. It is never a good strategy to try to answer, say, four out of five questions on a paper. If these were equally weighted questions, then you could only score a maximum of 80%.

Paper 1

- This is a multiple-choice paper made up of 40 questions each worth 1 mark. The paper is worth 20% of the whole exam.

- The exam lasts one hour. Aim to spend no more than $1\frac{1}{2}$ minutes on each question.

- Questions on the paper are taken from any topics in the core (topics 1–8) or the additional higher level (AHL) parts of the syllabus (topics 9–14). There are roughly 20 questions from the core and 20 questions from AHL.

- The questions are arranged in syllabus order (roughly the order of the chapters in this book).

- There is no reading time allowed.

- You are not allowed a calculator.

- If English is not your first language, you can use a simple translation dictionary.

- You mark your answers **in HB pencil** on an answer sheet. If you decide to change your mind later in the test, you need to able to rub out and replace the original mark.

- If you cannot answer a question straight away, **leave it and move on to the next**. Everyone has different strengths in physics, and some of the later questions may be much easier for you. But if you do not get that far, you will never know! It is helpful to mark the question (not the answer sheet) to remind yourself that you have left it until later.

- Some students like to write down the answer code for each question on the paper too. This means that, at the end, you can check that, if you thought the answer was "C", you actually wrote down "C" on the answer sheet.

- Try to leave time to check your answers. Finishing the paper has priority, but when you are practising before the exam, try to get through the whole paper in 55 minutes to allow for this.

- Paper 1 only tests at the level of objectives 1 and 2, not objective 3. (You can find out more about objectives in chapter 3.)

Paper 2

- Like paper 1, this paper tests the whole of the core and the AHL parts of the syllabus.

- The paper lasts 2 hours 15 minutes and is worth 36% of the whole exam.

- You answer on the paper itself.

- Paper 2 tests all three objectives. The paper is designed so that there are roughly the same number of marks allocated to objective 3 as to the other two combined.

- There are two sections to the paper: A and B.

- Section A is compulsory—you answer everything. The section has a total of 45 marks and will always contain a data-based question in which you analyse a set of data. These data may be in the form of a table or a graph. Chapter 6 gives you an example of a data-based question, and some advice on how to answer it. The data-based question usually accounts for about 10–12 marks. The remaining marks will be spread among four or more other questions and will cover a broad (although obviously not complete) range of the syllabus.

- Section B contains four questions (B1–B4). You are required to answer **two** of these. Each question has 25 marks and is known as an extended-response question. This means that each question examines one or two areas of the syllabus in particular depth. A look at the final chapter of this book will show you the style.

- You need to make a sensible choice of questions from section B. A poor strategy is to try to "spot" syllabus areas before the exam and only revise these. A much better strategy is to revise thoroughly for your exam and then choose your questions carefully.

- An effective choice begins during your reading time, with a consideration of the *optional* questions. Do not, initially, spend reading time on section A. Read all the B questions carefully and make a decision as to which questions suit

you best. Are you better at calculations than writing? Are you usually better at one syllabus area than another? Are the topics combined in a way that is better for you in some questions rather than others? Only when you are completely comfortable with your choice of B questions should you begin to look at section A during the reading time.

- Paper 2 exams do not need to be answered in question order. Always begin with the questions that you find most straightforward. Then return to the ones that you think are the more difficult, for whatever reason.

Paper 3

- This paper tests the options part of the syllabus.

- The paper is timed at 1 hour 15 minutes. This paper is worth 20% of the whole exam.

- As with Paper 2, you answer on the paper itself.

- This paper also addresses all three assessment objectives on the same basis as in paper 2: objectives 1 + 2 taken together have the same weighting as objective 3.

- Within an option, each question is compulsory, so there are no decisions about questions to be made.

- You will know which options you have studied. Take care to answer the correct ones. Every year it is clear to examiners that some students have answered the wrong option, perhaps by error, perhaps because they thought that the wrong option was easier than the one they should have answered. Remember that the option questions go into the physics in much greater depth than do paper 2 questions on apparently similar topics. Students answering the wrong option will not appreciate some of the subtleties of the physics and are likely to score poorly in this paper.

- Each option will contain several short-answer questions (such as those in paper 2, section A) and one extended-response question (more like those of paper 2, section B). However, the differences between the two types of question are less marked than in paper 2.

Coversheets and stationery

You will be provided with a personalized coversheet that you will attach to your paper 2 and 3 scripts. You will have a personalized computer sheet for paper 1.

At the start of the exam, check that all your details on the coversheets are correct. If not, inform the invigilator straight away. It is important to complete the coversheet correctly at the end of the exam. Enter the sections or options and the questions answered.

You should receive paper for rough work for every paper. Remember that this is not returned to the examiner, so be careful that you do not write important parts of answers on it. In addition, you should receive or will need the following things.

For **paper 1**:

- the personalized answer sheet from the IBO
- the question paper
- a data booklet.

For **papers 2 and 3**:

- the personalized coversheet from the IBO
- the answer book that contains the questions
- a data booklet
- additional lined answer sheets if you need them
- a sheet of graph paper (ask for more if you make a mistake)
- a treasury tag to fix all the papers, the coversheet and the graph paper together.

You can use a graphic display calculator (but not in paper 1) and simple drawing instruments (ruler, protractor, set square). You will want to ensure that you have a reliable set of pens (filled!) and pencils (sharp!).

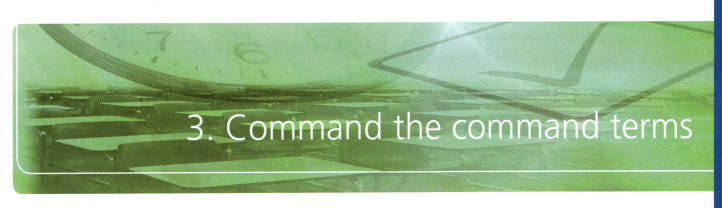

3. Command the command terms

This chapter will help you to understand the command terms that are used in exam papers. These terms tell you about the depth your answers should go into and how they might be structured.

IB physics questions are asked using a very limited set of instructions such as "**state**", "**explain**", "**calculate**" and so on. These clear instructions (called "command terms") tell you how to answer the questions.

Command terms show what IB examiners expect a student to be able to achieve at the end of the course. Science is a series of linked concepts and facts, and it has its own unique set of techniques. Science terminology is specialized—facts, methods and concepts are presented in an agreed way. You will demonstrate your mastery of these areas in the exam within three groups that are known as "objectives". The Diploma Programme *Physics guide* tells you the objective level of a topic, as the command term is used to specify what you need to know.

The first group of command terms is designed to allow you to show your basic competence in demonstrating your knowledge of methods, concepts, facts and so on. The second group of terms tests your ability to apply this knowledge, while the third group allows you to show your ability to analyse and evaluate scientific ideas.

Assessment objective group 1

These terms test your factual knowledge. Examiners will want to see an ability to recall standard facts and ideas.

Define
A question such as
"**Define** specific heat capacity."
requires you to give a clear answer that specific heat capacity is the energy required to increase the temperature of one kilogram of a substance by one kelvin. This is the standard textbook definition, and is clearly identified in the IB physics syllabus.
Many students find it worthwhile to memorize all the definitions from the syllabus.

Draw
and
Label
These often go together. You may be asked to
"**Draw** the *I–V* characteristic for a filament lamp."
or
"On the diagram, **draw** and **label** the forces acting on a mass on a slope."
Even though the question may be a simple one involving recall, take care in presenting your answer. Draw the diagram carefully. Add labels to the diagram neatly, making it clear which part of the diagram you are labelling.

State
This is used when the examiner requires a brief answer without any explanation. For example,
"**State** the unit of resistance."
requires the answer "ohm" but does not require an statement of what the ohm means.

List
and
Measure
These are used only rarely in physics exams. Their meanings, when they do appear, are obvious.

Assessment objective group 2

The second group of command terms requires you to apply your knowledge.

Calculate

Here you use data—given to you in the question itself or in the *Physics data booklet*—to find a numerical answer from a calculation. For example:

"An object is dropped from rest near the surface of the Moon. The acceleration due to gravity at the surface of the Moon is $1.6\,m\,s^{-2}$. **Calculate** the speed of the object $2.5\,s$ after release."

A good answer to a "calculate" question contains

• the use of a relevant equation (not just quoting it)

• any assumptions made in carrying out the calculation

• the substitution of appropriate data into the equation

• an accurate numerical calculation with an answer expressed in correct form with an appropriate unit.

Describe

A familiar term that covers more options than "**state**". Examples of the use of "**describe**" in the Diploma Programme *Physics guide* are:

"**Describe** the evidence that links global warming to increased levels of greenhouse gases."

"**Describe** the interchange between kinetic energy and potential energy during SHM."

In each case you are required to give a *detailed* account of the topic.

Before you begin to write, take a moment to plan your answer. Ask yourself the following.

• Have you included all the points that the examiner is likely to want?

• Have you matched the number of different points in your answer to the number of marks available for the answer (printed in the right-hand margin)?

• Is your planned answer in an order that is logical and non-repetitive?

Estimate

This command term has a more specialist meaning and needs care. An "**estimate**" in physics is an approximate answer to a problem where an exact value may never be known. A typical "**estimate**" question is:

"**Estimate** the number of words in this book."

It is possible to count every word in the book, but no one really needs such detailed information. A better way is to realize that estimate of total number of words = (average number of words on a line) × (average number of lines on a page) × (number of pages in book).

Identify

This instructs you to find an answer from a number of possibilities. An example from a paper 2 is:

"Argon-39 undergoes β^- decay to an isotope of potassium (K). The nuclear reaction equation for this decay is

$${}^{39}_{18}Ar \rightarrow K + \beta^- + x$$

State the proton (atomic) number and the nucleon (mass) number of the potassium nucleus and **identify** the particle x."

[3]

This example also shows you how command terms can sometimes be combined in one question. There are other examples of this later.

Outline	An "**outline**" is a brief account of a piece of physics or a summary of the points in an argument. For example:
	"During the night, the air temperature of a frozen lake drops to –20°C. Surface ice melted during the day freezes again. **Outline** one mechanism by which thermal energy is lost by the ice." [2]
	There are two points to make in the answer (because there are 2 marks) and a number of ways to answer. A typical outline answer might be that the water–ice surface is warmer than the surroundings (1 mark) and so it radiates electromagnetic waves, thus losing thermal energy to the surroundings (another mark). There is a description of the physical processes going on but no detailed explanation of them.
Distinguish	This is a rarely used term in physics exams that asks you to list the differences between a number of items or ideas.
Annotate and **Apply**	These two terms appear in group 4 exams but are only rarely used in physics.

Assessment objective group 3

The final group of command terms requires you to use high-order skills of analysis and evaluation in using your physics.

Explain	"**Explain**" may appear to be very similar to "**outline**" from assessment objective 2. The difference is that the requirement here is for a detailed account of the underlying physics. "**Explain**" tells you that more detail is needed than in an "**outline**". In the "**outline**" question above, had this been an "**explain**", then more marks would have been available. More detail would have been required—for example, the difference between thermal energy being lost and gained, or the nature of electromagnetic waves. Additionally, a good answer might have required reference to convection in the air leading to replacement of warm air with cold gas at the surface of the ice.
Discuss	This has a similar meaning to "**explain**". In this case you may be asked to compare alternatives, to give views for or against an argument, or to give the relative sizes or importance of factors. For example:
	"**Discuss**, in terms of the movement of the electrons, the energy transformations taking place in the filament of a lamp." [4]
	The question (which has 4 marks assigned to it) is looking for answers that deal with the reasons for changes in the kinetic energy of the electrons as they accelerate in the filament and lose energy to the filament atoms by collision. This is a high-level answer and as such demands that you think carefully before beginning to write.
Suggest	You should propose a hypothesis from the facts that you have available. For example:
	"The uncertainty of the radius is ±0.5 m. The addition of error bars to the data points shows that there is a systematic error in the plotted data. Suggest **one** reason for this systematic error." [2]
	The 2 marks here will be for the systematic error itself and for saying why the error leads to the addition of a constant value to each data point.

Show

and

Determine

These are two terms that have close parallels. They require you to go beyond the skills required for a calculation but, like "**calculate**", they both involve the manipulation of data to reach a numerical answer.

A "**determine**" question is a calculation for which full algebraic and numerical working is required. There will be 1 mark for the answer, and if you give only the correct answer with no working, 1 mark is all you will receive.

"**Show**" (often "**show that** …") is essentially a "**determine**" where you are given the final answer and asked to indicate the steps that lead from the question to the answer. This sounds easy! In fact "**show that** …" answers are often poor in physics exams. To gain **all** the marks you must give **all** the information that the examiner requires. The answer by itself gains nothing—after all, it was given to you.

In preparing for your exam, practise giving clear step-by-step solutions to both "**show that**" and "**determine**" questions.

"**Show that**" questions often lead to another calculation following on from the answer that you were asked to produce. Always begin the later parts using the "**show that**" value in the question, not your own if it was different. Do not cross out the part you could not solve properly (you may have gained some credit, as the examiner will allow credit for an early error), but remember to use the "**show that** …" value, **not** your own.

Comment

This command term requires a judgment based on a calculated numerical or other answer. If the answer differs from another quoted value in the paper, then you should try to work out why. Could errors have occurred in the particular case you were considering? Were there factors that might have changed the answer?

Compare

"**Compare**" asks you to look at more than one piece of information and to draw out similarities and differences. A common fault is to begin with the similarities and to forget to write about the differences.

Derive

and

Deduce

These can be treated as similar to "**show that**" in terms of the examiners' expectations of your answer. The terms are more likely to be used when the steps are to be carried through algebraically rather than numerically to a final equation, such as:

"**Deduce** that the wave power P per unit length of the wavefront is given by $P = \frac{1}{2}A^2Vg\rho$."

The guidance here is the same as for "**show**" and "**determine**": show every step, in the correct order, and with a full written explanation. Examiners can only base the mark on what you have written.

Sketch

At first sight you might imagine that "**sketch**" it is the same as "**draw**" (which is in assessment objective group 1). The level at which you need to answer a "**sketch**" question is significantly greater than for "**draw**". It is likely that you will need to take information from elsewhere in the question and add it to the graph or diagram. You will not usually be asked for straight recall in a "**sketch**" question—there will normally be some prior evaluation.

As a pointer to answering a "**sketch**" question on a graph, make sure that **axes are labelled** with appropriate **units** and that the **axes have numerical values** to indicate the scales. The "**sketch**" question may follow one or more numerical calculations that relate to the graph. If you have derived data that relate to the graph, make sure that the details of your graph agree with these data. Add further numerical data or actual data points to the graph if you can.

The following terms at assessment objective group 3 level are only rarely used in physics exams and are given here for completeness.

Analyse   Interpret data to reach a conclusion.

Construct   Represent or develop in graphical form.

Design   Produce a plan, simulation or model.

Evaluate   Assess limitations and implications.

Predict   Give an expected result.

Solve   Obtain an answer using algebraic and/or numerical methods.

Combining the command terms

Sometimes the command terms are grouped together. For example:

"**State** and **explain**, by reference to energy transformations, whether the speed with which the ball hits the ground is equal to $30\,\text{m s}^{-1}$.   *[2]*

There are two parts to this question: (1) does the ball hit the ground at $30\,\text{m s}^{-1}$ or not; and (2) explain how you reached your conclusion to (1). The important thing is to include both parts in your answer. The explanation can be first or second, but arrange your answer logically and make the statement clear.

The need for the answer to be clear and complete applies whenever there are two command terms together. Try to think of the parts as distinct. If you cannot answer the higher-order part of the question, make sure that you do as much as you can of the easier part.

Here are tips that will help you during the exams. Some points may seem obvious, but they are sometimes forgotten by students.

Paper 1

- Identify challenging questions during a read-through of the paper. Tackle easier questions first.

- For each question, read the stem (the question itself) carefully and try to arrive at an answer before reading the four alternatives.

- Underline or highlight key words.

- There can be important but easily missed words, such as "not" and "always".

- Next, read all of the alternatives carefully, even if the first choice seems correct.

- Eliminate choices you know cannot be correct.

- The most obvious answer is likely to be correct. Do not read too much into the question.

- If you are unsure, go with the most obvious answer.

- Do not forget that you can write on the question paper. It is sensible to write algebra or numbers in the blank space near a question rather than do the work in your head and risk a slip.

- When you have identified the best response, mark it carefully on the answer sheet in pencil.

- If you cannot answer a question, move on to the next. It is important to answer all the easier ones rather than run out of time by spending too long on a question you find difficult.

- If there is time, check that you have made the correct mark on the sheet for each question.

Paper 2 and paper 3

- Spend your reading time wisely. Your job here is only to select two of the four section B questions. Concentrate on making a sensible choice that plays to your strengths in physics.

- Begin by reading the whole of the question.

- You will be penalized for the omission of or an error in units. In your preparation for the exam, try to make the addition of a unit to a value completely automatic.

- You will be penalized for an inappropriate number of significant figures (often referred to as sig. fig., s.f. or SD) in a **final** answer. You should quote the same number of significant figures as were used in the question or no more than one extra. You will see examples in later chapters where students have been penalized for using incorrect numbers of significant figures. If the question asks you to "**show that**" an answer is a particular value, you are advised to quote a number of significant figures **greater** than the number used in the question to prove that you have actually carried out the calculation.

- Draw graphs and scale drawings carefully. Use a pencil (in case you make a mistake) and a ruler (to get accurate answers). You may be marked on your drawing accuracy.

- The number of marks quoted in the answer always gives the number of steps in an argument or points in a discussion.

- Write inside the space provided for the question—do not stray outside. Use additional sheets as sensibly as possible, and give the examiner a **clear** indication that your answer moves to an additional sheet.

- Neatly delete work that you do not want marked, but do not obliterate it. The examiner may find something worth marks under the crossing out.

Communication strategies

- Interpret the command terms correctly.

- State everything that is important, even if it seems obvious. This is particularly important if you make assumptions in a calculation.

- Do not waste time rewriting the question. Begin your answer with "This is because …".

- Do not quote your answers in the way that numbers appear on calculators or on computer spreadsheets, for example, not as 1.78E-8 or simply 2.36 -09. The correct way to express an answer is in the form: $3.00 \times 10^8 \, \text{m s}^{-1}$. Even $300\,000\,000 \, \text{m s}^{-1}$ looks awkward and implies a much greater level of precision than you might intend.

- The Diploma Programme *Physics guide* indicates that there are some equations you can be expected to derive. Do not learn these proofs—instead, learn the way the physics ideas are combined. You will understand the work much better this way.

- Certain types of equation recur many times in physics. An example is the exponential equation $y = y_0 e^{-\lambda t}$, which occurs in different forms in radioactivity, medical physics and a number of other places. Use all of these to reinforce your understanding of what equations tell us about the subject.

Drawing graphs

The straight-line graph is one of the important tools that the physicist uses to analyse data. It shows directly the relationship between two variables, called *y* and *x* in the general case (but in the real case labelled by their names, such as speed and time).

The straight-line graph has two important elements, the gradient *m* and the intercept *c*, as shown in the diagram.

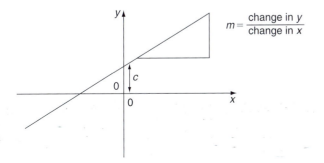

$$m = \frac{\text{change in } y}{\text{change in } x}$$

You may be given a false origin on the exam paper (where the bottom left corner is not $(0,0)$). Take care with a graph like this, especially if you have to determine an intercept on one of the axes.

- Remember the rules for calculating the gradient of a straight-line graph.

- Draw your triangle for the gradient on the graph grid so that the examiner knows what you are doing.

- Make the triangle large—with its largest side at least half the graph line.

- Check and double check your read-offs, including powers of 10, from the axis where necessary.

- Set out your calculations clearly and logically.

- Give the units of your answer.

- Be prepared to sketch the shape of common graphs: sine and cosine, $\frac{1}{x}$, $\frac{1}{x^2}$ and so on.

Using the *Physics data booklet*

- Know where everything is in the booklet. Do not waste time finding your way round it in the exam.

- Learn the most common equations (for example, equations of motion, electrical equations, $V = IR$ and so on).

- When asked to derive a constant, do not assume that the numbers in the question will give exactly the accepted answer as specified in the booklet.

Doing calculations

- Explain all calculations, even the simplest. If you go wrong, the examiner can look through your work to give partial credit.

- Show all your substitutions—there are usually marks for this.

- You will not usually get marks simply for copying an equation from the *Physics data booklet*. You will gain marks for rewriting it in the form that is required.

5. Key advice on revision

This chapter indicates ways in which you might tackle revision for HL physics exams. You can supplement your supply of past papers and markschemes at the IB store at http://store.ibo.org.

Revision strategies

- Have the tools for the job to hand. To revise effectively, you will need the relevant pages of the Diploma Programme *Physics guide*, your physics notes and a course textbook. Pens, marker pens and stationery are obviously essential.

- Revision should be active, not passive. Re-work your notes, add marginal notes to notes and texts, answer questions, have quiz sessions with your friends. Simply reading passively through a textbook or a set of notes does not make for good revision. Make your brain re-work the material.

- Build more complex material on a secure knowledge of basic ideas. Start your revision from the beginning and keep testing your understanding.

- Have a revision timetable. Start your revision early—six months before the exams is not too soon. Aim to go through each topic at least twice. Make sure that you devote sufficient time to revision of the two options.

- Do not spend too long on one area of study. Most people work effectively in bursts of about 20–30 minutes. Do not make the overall revision session too short. An ideal study session would be 2–3 hours long.

- Try to make revision fun. Give yourself rewards for time spent on revision or for good marks gained in class tests. It does work!

- Multiple-choice past papers are an excellent way of testing your knowledge and determining your strengths in the subject.

- Compare answers you write for papers 2 and 3 with the published markschemes.

- One way to answer past papers is to:
  - read the question, and decide which (if any) parts of it you are going to find difficult
  - study the relevant section again until you are able to answer the question
  - shut any notes and texts before answering the question under exam conditions
  - check your answers against the markscheme.

- Towards the end of your revision, answer some complete papers against the clock.

Interpreting markschemes

Markschemes are written for examiners, not students, but you can gain a good deal from studying them. Here is part of a paper 2 question with the associated markscheme.

Sand falls vertically on to a broad horizontal conveyor belt at a rate of $60\,kg\,s^{-1}$. The conveyor belt moves with speed $2.0\,m\,s^{-1}$. When the sand hits the conveyor belt, its horizontal speed is zero.

(a) Identify the force F that accelerates the sand to the speed of the conveyor belt. *[1]*

(b) Determine the magnitude of the force F. *[2]*

(c) Calculate the power P required to move the conveyor belt at constant speed. *[1]*

Markscheme

(a) friction / resistance between sand and belt ; *[1]*

(b) recognize that F = rate of change of momentum ; *[1]*

$$\left(\frac{\Delta m}{\Delta t}v\right) = (60 \times 2.0) = 120\,N;$$ *[2]*

(c) $P = 120 \times 2.0 = 240\,W;$ *[1]*

- Each marking point begins on a separate line. The end of the marking point is shown using a semicolon (;).

- An alternative answer or wording is indicated in the markscheme by a "/" (see part (a)). Either wording can be accepted. Examiners will not look for the exact words quoted in the markscheme but will allow the idea however expressed.

- Underlined words or phrases are, however, compulsory and the mark will not be given unless the word, phrase or idea is present (see part (b)).

- Words in brackets (…) in the markscheme are not required to gain the mark, but are used to clarify matters for the examiners.

- The order of points does not have to be as written (unless stated otherwise).

- Marking is always positive. Answers are given credit for what they achieve and for correct ideas.

- Sometimes a range of acceptable answers is allowed and the markscheme will have an associated uncertainty figure, such as 41 ± 2. This is a note to the examiner to accept answers between 39 and 43. When writing down your numerical values to exam questions, it is not necessary to include an uncertainty figure except sometimes in question A1, the data-analysis question.

6. Physics and physical measurement

Key terms for this chapter

- fundamental units and derived units
- random errors and systematic errors
- precision
- accuracy
- absolute, fractional and percentage uncertainties
- vector and scalar

The realm of physics

You should know:

- the ranges of magnitude for masses, distances and times in the universe.

You should be able to:

- state quantities as orders of magnitude or as differences of orders of magnitude
- estimate the approximate orders of magnitude for everyday quantities.

Example

Which of the following, in metres, is the order of magnitude of the diameter of the nucleus of a hydrogen atom?

A. 10^{-10} m

B. 10^{-15} m

C. 10^{-23} m

D. 10^{-30} m

Answer: B

It is always important to read the question carefully. It asks for the nuclear diameter, not the atomic diameter. One of the wrong answers is the diameter of the atom itself, 10^{-10} m. The two smallest answers relate to objects much smaller than the nucleus. You should have a reasonable knowledge of the relative and absolute sizes of the atoms, molecules and atomic particles.

Be prepared

- The Diploma Programme *Physics guide* indicates the ranges that you should consider for the masses, distances and times. Ensure that you have learnt these.

- Estimates are "educated guesses". There is an example of a possible estimate in the section of chapter 3 that deals with this command term. As a general rule, leave your answer as an order of magnitude (the nearest power of 10) unless the question specifies a number of significant figures.

Measurement and uncertainty

You should know:

- the following fundamental units in the SI system and their abbreviations—kilogram (kg), metre (m), second (s), ampere (A), mole (mol), kelvin (K)
- the difference between fundamental and derived units
- the difference between precision and accuracy
- what is meant by random error and by systematic error.

You should be able to:

- state units in the accepted SI format
- give examples of derived units and convert derived units
- state values in scientific notation
- use appropriate prefixes
- calculate results to an appropriate number of significant figures
- determine uncertainties in results
- state and calculate uncertainties as absolute, fractional and percentage uncertainties
- interpret uncertainty in results as error bars on graphs and vice versa
- calculate the uncertainties in the gradient and intercept of a straight-line graph.

Example

An object falls from rest with an acceleration g. The variation with time t of the displacement s of the object is given by

$$s = \tfrac{1}{2}gt^2$$

The uncertainty in the value of t is ±6% and the uncertainty in the value of g is ±4%. The best estimate for the uncertainty of the position of the object is

A. ±5%

B. ±8%

C. ±10%

D. ±16%

Answer: D

Think of the calculation of s as 0.5 × g × t × t. The 0.5 has no uncertainty, so as this is a multiplication, add ±4% to 2 × (±6%).

Be prepared

You are always expected to be able to express your answers with the correct unit and, where necessary, give the correct prefix (for example, m for milli, k for kilo, and so on). There is a list in the *Physics data booklet*.

- **Absolute uncertainty** is the size of an error (and its unit)—for example, a one-metre ruler graduated in millimetres can be read to the nearest 0.5 mm.
- **Fractional uncertainty** is given by

$$\frac{\text{absolute uncertainty}}{\text{magnitude of measurement}}$$

- **Percentage uncertainty** is 100 × fractional uncertainty.
- Combining errors is something else that you need to practise.

Vectors and scalars

You should know:

- the difference between vectors and scalars.

You should be able to:

- give examples of vector and scalar quantities
- add and subtract vectors using a graphical method
- resolve vectors into components at 90° to each other.

Example

The diagram below shows two vectors, x and y.

Which of the vectors below best represents the vector c that would satisfy the relation $c = x + y$?

A.

B.

C.

D.

Answer: B

You are asked to find the sum of $x + y$. To solve this, sketch on the question paper (or in your mind's eye) vector x and then from the arrow head (the point where the vector has reached) sketch y in the correct direction and at about the correct length.

Be prepared

- If you are asked to calculate a vector quantity, then your answer must always indicate the **direction** as well as the size and the unit. This will apply throughout the exam papers, so a mechanics question that asks for a velocity (a vector) will require the direction in which the object is moving.

- Be ready to draw scale vector diagrams. State the scale you are using. Use a ruler and where necessary a protractor. Practise this skill carefully.

A1. This question is about liquid flow.

The diagram shows a storage container for liquids.

The container is filled from above. The distance between the base of the container and the ground is h_0.

The container, which is initially empty, is then filled at a **constant** rate. The height h of the liquid surface above the ground is measured as a function of time t. The results of the measurements are shown plotted below.

liquid surface height above ground h / m

time t / s

(a)   Draw a best-fit line for the data.   *[1]*

(b)   It is hypothesized that h is directly proportional to t. State and explain whether this hypothesis is correct for the periods

    (i) $t=0$ to $t=120\,s$.   *[1]*

    (ii) $t>120\,s$.   *[1]*

(c)   Use data from the graph to determine the value of h_0.   *[2]*

(d)   The area of the base of the container is $1.8\,m^2$. Deduce that the volume of liquid entering the storage container each second is approximately $0.02\,m^3\,s^{-1}$.   *[3]*

(e)   The container is completely filled after $850\,s$. Calculate the total volume of the container.   *[1]*

(f)   The empty container is now filled at half the rate in (d). Using the axes, sketch a graph
       to show the variation of h with t in the range $t=0$ to $t=900\,s$.   *[2]*

[Taken from HL paper 2, time zone 2, May 2009]

How do I approach the question?

Begin by reading the whole of the question. Think about how the liquid height
will vary with time while the liquid is filling (a) the cylindrical section, and (b) the
conical section above the cylinder.

(a)   "Draw a best-fit line" means use a pencil and, where necessary, a ruler, to
       provide a line that minimizes the distances between points and the line.
       Try to get a balance of points on each side of your line. Notice that the use
       of the word "line" does **not** imply that the line has to be straight.

(b)   A hypothesis is a possible suggestion about a relationship that can be
       tested, in this case that h is directly proportional to t. If h is directly
       proportional to t, then the graph of h against t must be a straight line that
       goes through the origin $(0,0)$. Does it?

(c, d) Think carefully about what the graph tells you. The value of the intercept
       when $t = 0\,s$ has a meaning. The value of the initial gradient of the graph
       also gives you information about the liquid flow into the container and
       hence the total volume. Work out what these are.

(e)   Finally, remember that the command term "**sketch**" is a (high-level) objective
       group 3 instruction, which means more than simply "**draw**". Consider
       carefully what is already printed on the graph axes and what you need to
       add.

Which areas of the syllabus is this question taken from?

- The Diploma Programme
  Physics guide says that paper 2,
  section A will contain a data-
  based question that will require
  you to analyse a given set of
  data. These data may be in the
  form of a table or a graph.

- Many of the sections in physics
  and physical measurement are
  examined in the data-analysis
  question. There is likely to
  be a particular emphasis on
  measurement and uncertainties,
  which includes the analysis of
  graphs.

This answer achieved 4/11

This student realizes that the line has to pass through all the error bars.

Notice the misread here. It leads the student into error later. The examiner looks at everything you write or draw and will be concerned about these errors.

In a simple question such as this, only 1 mark covers the answer and its explanation.

The statement is correct, but for the mark the explanation must focus on the shape of the graph.

Because the graph line is drawn (incorrectly) to the (0, 3.5) point, the mark is allowed as an error carried forward. The explanation is correct too after this initial misunderstanding.

A calculation of the gradient is required here. The student thinks that a liquid height of 5.1 m has been filled in 120 s and calculates the rate of height change. This should then have been **multiplied** by the area of the base.

The total fill time is 850 s, the filling rate is 0.02 m³ s⁻¹ and so the total volume is the product of these.

To the examiner's eye the initial gradient is the same in both graphs. If the fill rate is half, the gradient should be half of the original

(a)

1

(b) (i) It is correct here, as the volume of the area being filled is constant.   0

(ii) It is incorrect here, as the shape of the container changes and so does the volume (the volume increases).   0

(c) h_0 must be 3.5 m, because initially the container is empty which means at $t = 0$ the surface of the water is equal to the bottom of the container; at 3.5 m.   2

(d) $\dfrac{5.1\ m}{120\ S} = 0.425\ m/s$ $\dfrac{0.425\ m/s}{1.8\ m^2} = 0.0236\ m^3\ s^{-3}$ 0

$\approx 0.02\ m^3 s^{-1}$

(e) $850\ s \times 0.02\ m^3 s^{-1} = 17\ m^3$ 1

(f)

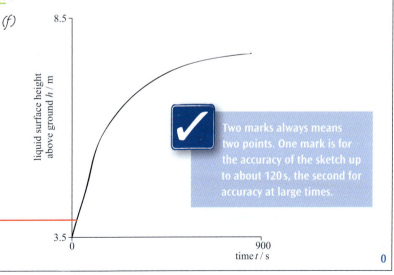

0

Two marks always means two points. One mark is for the accuracy of the sketch up to about 120 s, the second for accuracy at large times.

This answer achieved 9/11

The quality of the work here is not good. Straight lines are not drawn with a ruler.

The line goes through every error bar.

This answer says that the line is straight for small times (when the cylinder rather than the cone is being filled). "Directly proportional" needs a **straight line** to go **through the origin** at (0, 0). This line does not, so the answer is wrong.

Benefit of the doubt has been given here because a logarithmic increase would mean a curve. It would have been better to concentrate on describing the curved graph line at large times.

This is a good but brief answer that also gives an uncertainty.

Take care when reading graphs. The point is (120, 5.6), not 110 s as the student thinks. This inaccuracy loses a mark.

The final value at 850 s is too low but the intention is very clear and credit is given.

There is a clear attempt to show a halved gradient even though the slope is not quite right.

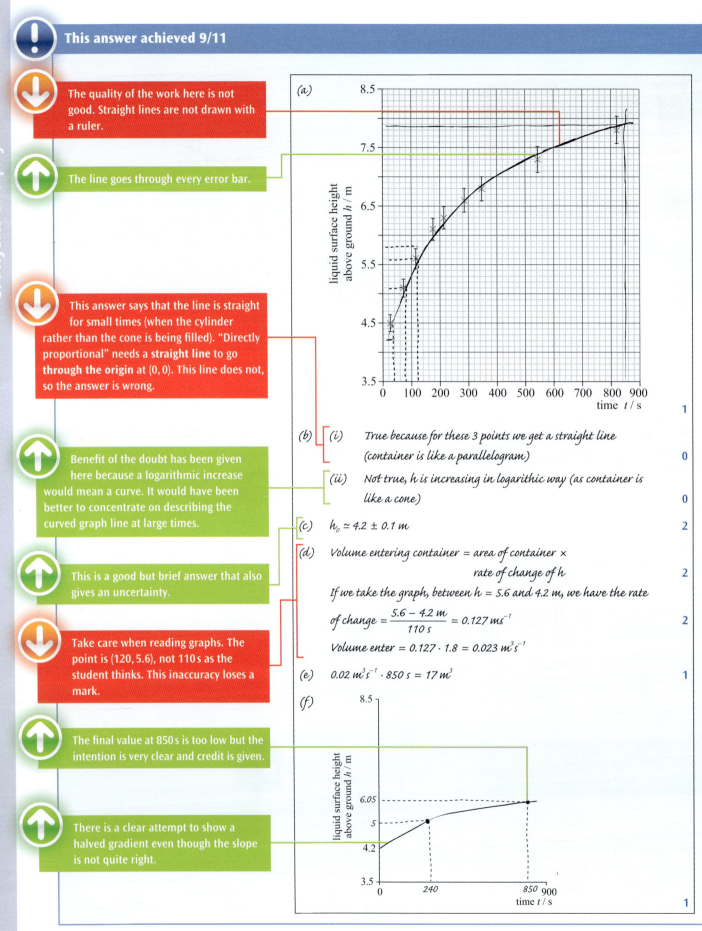

(a)

(b) (i) *True because for these 3 points we get a straight line (container is like a parallelogram)* — 0

(ii) *Not true, h is increasing in logarithmic way (as container is like a cone)* — 0

(c) $h_0 \simeq 4.2 \pm 0.1$ m — 2

(d) *Volume entering container = area of container × rate of change of h* — 2

If we take the graph, between h = 5.6 and 4.2 m, we have the rate of change $= \dfrac{5.6 - 4.2 \text{ m}}{110 \text{ s}} = 0.127 \text{ ms}^{-1}$ — 2

Volume enter $= 0.127 \cdot 1.8 = 0.023 \text{ m}^3\text{s}^{-1}$

(e) $0.02 \text{ m}^3\text{s}^{-1} \cdot 850 \text{ s} = 17 \text{ m}^3$ — 1

(f)

This answer achieved 10/11

The student's original line was rather thick above 200 s. As a rule, try to keep the drawn line thinner than the thickest printed grid line.

Unlike the previous example, this graph is drawn freehand but confidently. The line is within all the error bars (all that is required for the mark).

It is correct to say that the line does not go through the origin and therefore h is not directly proportional to t.

The answer is complete. Because the line is curved, it cannot show proportionality.

This is a good try that gets full marks, but the ideal answer shows the initial straight line extended (as shown below). This would enable a better estimate of the gradient.

(a)

1

(b)

(i) The hypothesis is not correct since the line does not go through the origin.

1

(ii) The hypothesis is not correct since the graph is a curve for $t > 120$ s.

1

(c) when $t = 0$ s,

$h = 4.2$ m

$\therefore$ $\boxed{h_0 = 4.2\ m}$

2

(d) $A = 1.8\ m^2$

when $t = 0$, $h = 4.2$.

$t = 120$, $h = 5.6$

$\Delta h = 1.4\ m$, $\Delta t = 120$

$\dfrac{v}{t} = \dfrac{1.4 \times 1.8}{120} = 0.021\Delta$

$\approx 0.02\ m^3\ s^{-1}$

3

This answer achieved 10/11 (continued)

This is not ideal. No marks are lost here, but it is better to put the unit close to the numerical result. There is always the danger of forgetting it.

A mark lost because the graph begins at 3.5 m. It should start at 4.2 m. The part of the graph for $t > 240$ s is fine, however.

(e) $0.850 \times 0.02 = 17$

$\boxed{17 \ m^3}$

1

(f)

1

7. Mechanics and motion in fields

Key terms for this chapter

- displacement, velocity, speed and acceleration
- instantaneous—at one moment in time
- average—mean value over time
- free-body diagram—a diagram showing all the forces that act on one object
- translational equilibrium—a particle in translational equilibrium has the sum of all forces = 0
- momentum—the vector quantity mass × velocity
- impulse—contact force acting on an object × time for which force acts
- power—the rate at which energy is converted
- efficiency—the ratio of useful energy output by a system to the total energy involved
- centripetal force—the force that keeps an object moving in a circle

Kinematics

You should know:
- that uniform accelerated motion is motion where the acceleration is constant with straight-line motion.

You should be able to:
- define and use the terms displacement, speed, velocity and acceleration
- use the kinematic equations for uniformly accelerated motion in a straight line:

  $v = u + at$ (not given in the *Physics data booklet*)

  $s = ut + \frac{1}{2}at^2$

  $v^2 = u^2 + 2as$

  $s = \frac{(u + v)t}{2}$

- draw and analyse graphs of motion:
  - distance–time and displacement–time
  - speed–time and velocity–time
  - acceleration–time
- calculate and interpret:
  - gradients of distance–time and velocity–time graphs
  - areas under velocity–time and acceleration–time graphs
- solve problems that involve relative velocity
- describe how air resistance acting on falling objects leads to terminal speed.

Example

The graph is a speed versus time graph for an object that is moving in a straight line.

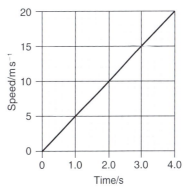

Kinematics (continued)

Which of the following is the distance travelled by the object during the first 4.0 seconds?

A. 80 m

B. 40 m

C. 20 m

D. 5 m

Answer: B

The area under the speed–time graph for a moving object gives you the distance travelled. Work out the area either:

* *with a calculation if the shape of the area is simple (as here)*

  or

* *by counting squares and by multiplying the number of squares by the distance equivalent of one square (in this example, the large squares are 1.0 s × 5.0 m s⁻¹, so 5.0 m).*

The trick is to use squares that are not so small that the count takes a long time, but not so large that the count is inaccurate. Mark the squares as you count them, so as not to count one twice.

Be prepared

* The kinematic equations only apply when the acceleration is uniform. If this is not the case, you will need to work from curved graphs of the motion in question.

* Be clear about the distinctions between distance and displacement, and between speed and velocity. Take care to use the terms correctly.

* Although three of the kinematic equations are given in the *Physics data booklet*, they are worth memorizing, as they are simple and familiar.

* Relative velocity problems are best solved using a vector approach. Remember to draw the vectors to scale.

* Graphs of speed–time and distance–time for an object falling through the air need care (see the diagram below). The object can fall from rest, or the object can start to move with a speed initially greater than terminal speed (for example, a bullet from a gun).

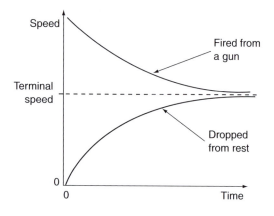

Forces and dynamics

You should know:

* the condition for translational equilibrium
* Newton's three laws of motion
* that momentum is mass × velocity and is a vector
* that impulse is force × time and is equal to the change in momentum
* that momentum is always conserved unless an external force acts on the system.

You should be able to:

* draw free-body diagrams
* calculate the weight of an object
* solve problems using Newton's laws of motion
* interpret the meaning of a force–time graph
* solve problems involving linear momentum and impulse.

Example

A lamp of weight W is suspended by a wire fixed to the ceiling. With reference to Newton's third law of motion, the force that is equal and opposite to W is the

A. tension in the wire.

B. force applied by the ceiling.

C. force exerted by the lamp on the Earth.

D. force exerted by the Earth on the lamp.

Answer: C

Newton's third law of motion gives rise to much confusion. In the example, four forces act, and these are the responses given. In fact, there are two pairs of forces, which are—for the purposes of Newton's third law—independent. If the wire breaks, then one of these two pairs disappears. But the other pair remains. The weight acting on the lamp (which is the force of the Earth on

Forces and dynamics (continued)

the lamp) causes the lamp to accelerate towards the centre of the Earth. The other half of the reaction pair also still acts. The Earth will accelerate towards the centre of mass of the lamp, but the Earth is so massive compared to the mass of the lamp that this will not be noticed.

Be prepared

- Free-body diagrams should show every force that acts on an object. There should be only one object in each diagram. The forces should be drawn in the correct direction and with the correct scaled length.

- Be able to write down correct statements of each of Newton's three laws and to describe a situation that

relates to a stationary or a moving object in terms of these laws (for example, how rockets or jet engines work, or how helicopters hover). Your explanation might need to include discussions of momentum transfer and impulse.

- Impulse is the product of force × time. The units are N s. This is the same unit as momentum, which can also be expressed as $kg\,m\,s^{-1}$. Impulse is equal to the change in momentum of an object. The area under a graph of force against time gives the total impulse that acts on an object during the time shown.

- Momentum is always conserved when no external forces act on a system. There are no exceptions to this.

Work, energy and power

You should know:

- work is the amount of energy transferred to an object when a force acts on it and the object moves as a result

- the direction between the force and the direction of the object changes the amount of work transferred to the object

- kinetic energy is the energy that a moving object has because of its movement

- gravitational potential energy is the energy a mass has because of its position relative to another mass

- the principle of conservation of energy

- elastic collisions do not lose kinetic energy

- inelastic collisions lose kinetic energy to other forms

- $\text{efficiency} = \dfrac{\text{useful power or energy taken from a system}}{\text{total power or energy given to a system}}$

- power is the energy transfer per second.

You should be able to:

- carry out calculations involving:
  - work
  - kinetic energy
  - gravitational potential energy
  - power
- describe energy transformations.

Example

A lift (elevator) is operated by an electric motor. It moves between the 10th floor and the 2nd floor at a

constant speed. One main energy transformation during this journey is

A. gravitational energy → kinetic energy.

B. electrical energy → kinetic energy.

C. kinetic energy → thermal energy.

D. electrical energy → thermal energy.

Answer: D

The elevator is moving at a constant speed, so there can be no change in the kinetic energy between the 10th and 2nd floor. Look carefully at the responses. How many does this eliminate? In what form does the electrical energy given to the motor ultimately end?

Be prepared

- Take care with units in all calculations involving work, energy and power. GW and MJ, and time units other than seconds, are all common in this area of the syllabus.

- As with the kinematic equations, the equations in energy and power are so common and straightforward that you are advised to learn them to avoid having to check the *Physics data booklet* during the exam.

- Work transferred to an object = force × distance **only** if the force and distance act in the **same direction**. If there is an angle θ between force and distance, then the equation becomes force × distance × cos θ. In fact, if the force and distance have an angle of 90° between them, then the work transferred is zero—which is why no work is done by a centripetal force.

Work, energy and power (continued)

- There are three types of "collision" (a collision includes any interaction between objects where energy and/or momentum are transferred).
  - Elastic (no change)
  - Inelastic (energy lost)
  - Superelastic (energy gained).
- Like momentum, energy is conserved. This refers to all forms of energy, including the energy E that Einstein tells us is in the form of mass m ($E = mc^2$), where c is the speed of light in a vacuum. In applying energy conservation, it is important to consider all the energy forms possible.
- Efficiency calculations need care. Include all the energy input to the system; useful energy taken out is usually more obvious. Remember that efficiency can be expressed as a percentage or as a decimal (between 0 and 1). Efficiency has no units.

Projectile motion

You should know:

- the horizontal and vertical components of velocity for a particle in a uniform field are independent of each other
- the path of a mass moving in a uniform field is a parabola if there is no drag or air resistance
- how the path of the moving object changes if there is air resistance (drag).

You should be able to:

- solve problems involving projectile motion
- use the kinematic equations
- use conservation of energy, which is sometimes easier than using the kinematic equations.

Example

An object moves in the x–y plane. The graphs below show how the component of its velocity V in the x-direction (V_x) and the component of its velocity V in the y-direction (V_y) vary with time t.

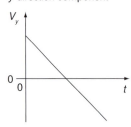

x-direction component / y-direction component

The particle is moving

A. in a parabola.
B. with simple harmonic motion.
C. with constant velocity in a straight line.
D. in a circle.

Answer: A

The speed in the x-direction is constant. The speed in the y-direction begins at a high positive value, dropping linearly to zero before increasing speed again in the opposite direction. What does this reversal of sign imply? Is the acceleration in the y-direction constant—the gradient will tell you this? What common motion does this represent, and, finally, what is the shape of this common motion?

Be prepared

- In a uniform gravitational field, the horizontal and vertical components of velocity are independent. The vertical component is the component that is in the same direction as the gravitational field strength. The gravitational field will accelerate the falling object in free fall. There is no component of acceleration in the horizontal direction.
- In the absence of air resistance or other frictional force, the horizontal component of speed will not change. The motion of such a projectile is parabolic.
- In calculations, split the horizontal motion from the vertical motion by taking components horizontally and vertically. Treat the vertical motion simply as that of an object falling vertically with an acceleration of g. The time of flight will enable you to calculate the range (the horizontal distance covered). You will then be able to recombine the horizontal and vertical components in order to establish the final velocity (do not forget to include the direction).
- Acceleration in **any** uniform field leads to similar ideas. An obvious example is a charged particle (such as an electron) that moves into a uniform electric field. The electron undergoes uniform acceleration and moves in a parabolic path.
- When dealing with objects falling vertically or which are projected, it can be easier to use an approach involving energy conservation. In a gravitational field and in the absence of frictional forces, the gravitational potential energy lost during the motion is equal to the kinetic energy gained ($\frac{1}{2}mv^2 = mgh$).

Uniform circular motion

You should know:

- an object always requires a centripetal force to make it move in a circular path
- the centripetal force is provided by the resultant of the forces acting on the system.

You should be able to:

- describe the origin of the centripetal force in various situations
- solve calculations involving centripetal acceleration.

Example

A particle is moving around the circumference of a horizontal circle of radius r with constant speed v. Which of the following is the acceleration of the particle?

A. $\dfrac{v^2}{r}$ towards the centre of the circle

B. $\dfrac{v^2}{r}$ away from the centre of the circle

C. $v^2 r$ towards the centre of the circle

D. $v^2 r$ away from the centre of the circle

Answer: A

The equation for centripetal acceleration is given in the Physics data booklet in two forms, one that involves v (speed) and r (radius of circle), and one in which v is replaced by T (time to go round the circle once). This is an equation that you might choose to memorize. The other decision for you is to choose the direction in which the acceleration acts.

Be prepared

- A centripetal force is always required to maintain the motion of an object in a circle at constant speed. The centripetal force acts inwards and is directed to the centre of the circle.
- Situations in which you might be asked to consider the origin of a centripetal force include:
  - planets around a Sun, or satellites around a planet
  - charged particles moving in magnetic fields
  - mechanics problems where a vehicle moves in a circle or where an object is tethered to a rope while moving in a circle.

Orbital motion

You should know:

- that gravity supplies the centripetal force that keeps a planet in orbit.

You should be able to:

- derive Kepler's third law
- derive the equations, for an orbiting satellite, of its:
  - kinetic energy
  - gravitational potential energy
  - total energy
- discuss what is meant by "weightlessness" in orbital motion, free fall, or in deep space.

Example

A communication satellite is moving at a constant speed in a circular orbit around Earth. At any given instant in time, the resultant force on the satellite is

A. zero.

B. equal to the gravitational force on the satellite.

C. equal to the vector sum of the gravitational force on the satellite and the centripetal force.

D. equal to the force exerted by the satellite's rockets.

Answer: B

This is a question that connects the physics of circular motion and orbital motion. If the resultant force on the satellite is zero, what does Newton's first law say about the motion? Is this possible in terms of the constant magnitude of acceleration that must act on any orbiting object? The rockets of the satellite are not required to keep it in a constant orbit at a constant speed. They would only be required if the orbital radius had to be changed. The gravitational force on a satellite supplies the centripetal force necessary to keep it in orbit, so can the two forces be represented as having a vector sum?

Be prepared

- Kepler's third law gives the relationship between the orbital time T and the orbital radius r of a satellite travelling in a circular orbit: $r^3 \propto T^2$. The way to remember the derivation is to recall that the gravitational force on the satellite is equal to the centripetal force needed to keep it at the required orbital distance.

- IB exams usually assume that the orbits of planets around the Sun and satellites around planets are circular (most solar system planets have orbits close to perfect circles). Some objects (for example, comets) have orbits that are long thin ellipses. There is a trade-off here between the kinetic energy and the gravitational potential energy. With

no energy losses (air resistance, collisions with meteors and so on), the total energy will remain constant. But as the distance of the comet from the Sun increases (r goes up), the gravitational potential energy increases. This means that the kinetic energy must become smaller to compensate for this change.

- You should be aware of the different interpretations of weightlessness: true weightlessness when far from any other mass, or the weightlessness of an astronaut orbiting the Earth.

A2. This question is about impulse.

(a) A net force of magnitude F acts on a body. Define the *impulse I* of the force. [1]

(b) A ball of mass 0.0750 kg is travelling horizontally with a speed of 2.20 m s^{-1}. It strikes a vertical wall and rebounds horizontally.

ball mass
0.0750 kg

2.20 m s^{-1}

Due to the collision with the wall, 20% of the ball's initial kinetic energy is dissipated.

(i) Show that the ball rebounds from the wall with a speed of 1.97 m s^{-1}. [2]

(ii) Show that the impulse given to the ball by the wall is 0.313 N s. [2]

(c) The ball strikes the wall at time $t = 0$ and leaves the wall at time $t = T$.
The sketch graph shows how the force F that the wall exerts on the ball is assumed to vary with time t.

The time T is measured electronically to equal 0.0894 s.

Use the impulse given in (b)(ii) to estimate the average value of F. [4]

[Taken from SL paper 2, time zone 1, May 2009]

28

How do I approach the question?

(a) Provide the normal definition of impulse. This must be in terms of momentum and relate to the quantity *F*.

(b) (i) This is a "**show that**" question and a clear presentation of your solution is vital. One way is to evaluate 80% (0.80) of the initial kinetic energy. However, as the mass does not change, it is perfectly acceptable to calculate 80% of (initial speed)^2 and then take the square root of this value to give the final speed.

(ii) Evaluate the momentum change of the ball (taking account of the **velocity** not the **speed**).

(c) The area under the *F*–*t* graph is equal to momentum change, so use this area and the definition of impulse in part (a) to estimate *F*.

Which areas of the syllabus is this question taken from?

- This question is based on "Forces and dynamics" (topic 2.2).

This answer achieved 2/9

This is an attempt at an explanation, not a definition.

Kinetic energy is related to v^2 not to v. The student realizes that 1.97 is 0.9×2.2; this is because $0.9^2 = 0.81$!

There is credit for the recognition that impulse $= m\Delta v$. But the final speed is used, not the velocity change.

There is recognition that impulse is $F\Delta t$ and a calculation to support this.

The student has omitted some important points of physics in the solution, as well as missing the unit, which incurs a penalty.

(a) *Impulse I of the force is the measurment of the magnitude acting on the body.*   **0**

(b) (i) $\dfrac{x}{2.20}$ $\dfrac{20}{100} \rightarrow .44 = 1.76$

$= 80\% + 10\% = 90\%$ $\dfrac{90}{100} \cdot \dfrac{1.98}{2.20} \approx 1.97$

$E_{vect} + 9$ **0**

(ii) *Impulse* $= F \Delta t = m \Delta v$

$(2.11)(.0750)(1.97) = .313$ **1**

(c) $I = F \Delta T$ $.313 = F(.0894)$ **1**

$F = 3.5$

This answer achieved 6/9

This is an accurate definition with a clear statement of the symbols used.

All the points needed are here, including calculations of initial and final kinetic energies.

This "**show that**" could be very much clearer, because factors appear without explanation (0.2 for example). A "**show that**" needs to be better presented than this.

The student has the correct idea, but has forgotten that momentum is related to **vector** velocity not scalar speed. The momentum change, taking into account direction change, is 1.97 − (−2.20) = +4.17 m s^{-1}.

The student gains partial credit but has not distinguished between the average and maximum values of the force.

(a) Impulse can be described as the change in momentum or $F \Delta t = I$
where F = force t = time in seconds $\qquad$ **1**

(b) (i) $E_k = \frac{1}{2} m v_f^2$ $\qquad\qquad\qquad$ $E_k = \frac{1}{2}(.0750)(2.20)^2 = .1815$

$.1458 = \frac{1}{2} m v_f^2$ $\quad m = .0750$ $\quad (.1815)(.2) = .0363$

$\dfrac{.145}{.0750} = \frac{1}{2} v_f^2$ $\quad \sqrt{3.86} = \sqrt{v_f^2}$ $\quad .1815 - .0363 = .1452$

$1.9\overline{3}/(\frac{1}{2}) = v_f^2$ $\quad 1.96\ ms^{-1} \approx v_f$ $\qquad$ ↵ $\qquad$ **2**

(ii) impulse = mass × speed change

$= 0.750 \times (2.20 - 1.97)$

$= 0.17$????

(c) $I = F \Delta T$ $\qquad\qquad$ $\Delta T = 0.894\,s - 0\,s = 0.894\,s$

$.313\,N\,s = F\,(.0894\,s)$ $\qquad I = .313\,N\,s$ given

$\qquad\qquad\qquad\qquad$ $F = unknown$

$\dfrac{.313\,N\,\cancel{s}}{.0894\,\cancel{s}} = 3.50\,N$ $\qquad$ ↵ $\qquad$ **3**

If the maximum force is F_{max} and the average force is F_{ave}, then the area $= \frac{1}{2} F_{max} \Delta t$, which is the impulse. For this shape, $F_{ave} = \frac{1}{2} F_{max}$ and the maximum force is double the average.

This answer achieved 8/9

The definition is given and clarified correctly.

This is a clear "show that" answer. There can be no doubt what is happening here, with every step made clear.

This is not quite so clear, but the meaning of mass × velocity is present and the numbers are added. This is the correct physics.

Points that should have been made include: velocities are used, not signs; and the positive sign is really two self-cancelling negative signs.

The student has appreciated the distinction between average and maximum force in the context of the triangle shape to the graph. This would not always apply if the variation of *F* against *t* was different.

(a) Impulse is the change in momentum of an object/body:

$I = \Delta P = m \, \Delta v$

It is the product of force and change of time   1

(b) (i)

$$\tfrac{1}{2} \, mv^2$$

$kE = \tfrac{1}{2} (0.0750)(2.2)^2$ $.1452 = \tfrac{1}{2} (0.0750) \, v^2$

$= .1815$ $.2904 = (0.0750) \, v^2$

$= .1815(.8) = .1452$ $3.872 = v^2$

$v = 1.97 \, m/s$ 2

(ii)   Impulse $= \Delta P = m \, \Delta v$

$(0.0750)(2.2) + (0.0750)(1.97)$

$.165 \; + .14775 = 0.313 \, N \, s$ 2

(c)   Impulse $= F_{ave} \, \Delta t$

$$\frac{0.313}{.0894} = \frac{(.0894)}{.0894} \, F$$

$F_{max} = 2 \times F_{ave}$ because it's a right angle triangle so $F_{max} = 7.0 \, N$ 3

B1. This question is about dynamics and energy.

A bullet of mass 32 g is fired from a gun. The graph shows the variation of the force F on the bullet with time t as it travels along the barrel of the gun.

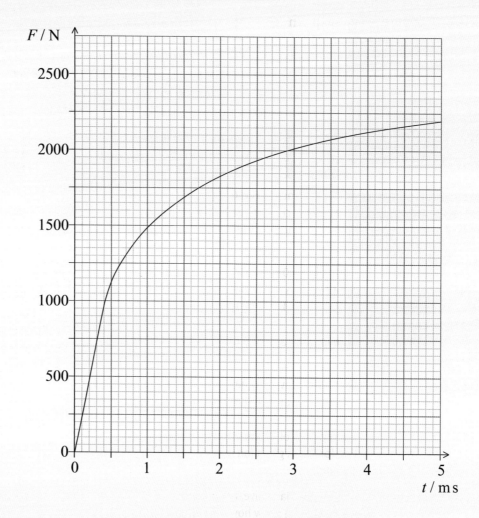

The bullet is fired at time $t=0$ and the length of the barrel is 0.70 m.

(a) State and explain why it is inappropriate to use the equation $s = ut + \frac{1}{2}at^2$ to calculate the acceleration of the bullet. *[2]*

(b) Use the graph to

    (i) determine the average acceleration of the bullet during the final 2.0 ms of the graph. *[2]*

    (ii) show that the change in momentum of the bullet, as the bullet travels along the length of the barrel, is approximately 9 N s. *[3]*

(c)   Use the answer in (b)(ii) to calculate the

    (i)   speed of the bullet as it leaves the barrel.   *[2]*

    (ii)   average power delivered to the bullet.   *[3]*

(d)   Use Newton's third law to explain why a gun will recoil when a bullet is fired.   *[3]*

[Taken from SL paper 2, time zone 2, May 2009]

How do I approach the question?

(a)   Force is always equal to mass × acceleration if the mass is constant (which it is here). Does the graph show that the force is constant with time? What does this suggest about the acceleration and hence the use of the kinematic equations?

(b)   (i)   The question asks you to focus on the last 2.0 ms of the graph. The graph line is approximately straight over this time range. Use this to estimate the average force and therefore, knowing the mass, the acceleration.

    (ii)   The area under the F–t graph is the impulse and also the change in momentum.

(c)   (i)   The initial speed of the bullet is zero. If the overall change in momentum and the mass of the bullet are known, then the speed can be calculated.

    (ii)   Power is the rate of transfer of energy. The total gain of kinetic energy may be calculated. The time taken for this is given (on the graph), and so the average power can be calculated.

(d)   Your answer must be in terms of Newton's third law. Give, as part of the answer, a clear statement of the law and go on to show how the gun's recoil arises as a result of the reaction force and hence an acceleration. Do not jump straight from a statement that there is a force to the suggestion that therefore there is movement.

Which areas of the syllabus is this question taken from?

- This question draws from a much broader part of the syllabus than the previous guiding question. As well as "Forces and dynamics" (topic 2.2), the question examines "Work, energy and power" (topic 2.3) directly. You will also need to understand elements of "Kinematics" (topic 2.1).

This answer achieved 3/15

No, we know quite a lot! Force is always directly proportional to acceleration, so this graph also gives the shape of an acceleration–time graph. But the curve shows that the acceleration is not constant.

This solution uses the **final value** of the acceleration, not the **average** acceleration over the last 2.0 ms.

The assumption is that the final force acts on the bullet for the entire 5.0 ms. This is not the case (as the graph shows clearly).

There is some credit here as the student has the correct idea what to do.

There is an incorrect use of the acceleration value from (b)(i) instead of the value of zero (the initial speed).

The solution is unclear. The value of 1.8 mN does not come from the data in the question.

There is no clear link between the A and B used in the first sentence.

(a) It is inappropriate because we know nothing about the acceleration of the bullet. **0**

(b) (i) $a = \dfrac{\Delta v}{\Delta t}$ $F = m\vec{a}$ $a = \dfrac{F}{m} = \dfrac{\Delta v}{\Delta t}$

$\dfrac{2200}{0.032} = \cdot 2 \cdot 10^{-3} = \Delta V$ average acceleration $=$ 137.5 m/s **0**

(ii) $\Delta p = F \Delta t$

$mv_2 - mv_1 = 2200 \cdot 5 \cdot 10^{-3}$

$\Delta p \approx 11$ not very accurate because the graph is not a straight line **0**

(c) (i) $mv_2 - mv_1 = 9 \ N \ s^{-1}$

$0.032 \ v_2 - 0.032 \cdot 137.5 = 9$

$v_2 = 418.75$ m/s **1**

(ii) power $= Fv$

$F = \dfrac{\Delta p}{\Delta t}$ $F = 1.8 \cdot 10^{-3}$

power $= 0.753$ W **0**

(d) Newton's 3rd law states that a force excerted on an object A is equal and opposit to the force on object B. So the force that is exerted on a gun when a bull is fired is opposit (in sign) and equal to the force exerted on the bullet. That is why the gun recoils. **2**

The student just gains credit for the idea that equal and opposite forces are exerted on bullet and gun.

This answer achieved 7/15

The idea of non-constant acceleration and hence the invalidity of the kinematic equations is present.

There is no link between non-constant force (on the graph) and non-constant acceleration.

The algebraic part of the solution is incorrect. The correct statement that $F = ma$ goes on to an incorrect statement that $a = I/m$, with I presumably being the impulse. The subsequent working does not receive credit.

The change in momentum is equal to the area under the graph. There is some further credit because the student has divided the area up into a rectangle and triangle, but the estimation is too crude and inaccurate for full credit.

There is not enough detail for any credit here.

This is an unusual method that gains most of the marks. The method is based on the equation: work done = force × distance.

The statement of Newton's third law is good, and there is a connection between the law and the equal and opposite forces, but there is no link to acceleration and the motion.

(a) The bullet is not accelerating at a constant rate, but the rate of acceleration decreases over time, as shown by the graph. The equation $s = ut + \frac{1}{2}at^2$ can only be used when acceleration is constant; therefore, it cannot be used to calculate the acceleration of the bullet.   **1**

(b) (i) At $t = 5 \times 10^{-3}$ s, $a = 440000 \ ms^{-2}$
        $F = 22000 \ N$ at $t = 3 \times 10^{-3}$ s,
    so, $F = ma$ $F = 2000 \ N$
        $a = \dfrac{I}{m}$ so, $a = 400000 \ ms^{-2}$

        $a = \dfrac{2200}{5 \times 10^{-3}}$ so, $a = 4400000 - 400000$
                            $a = 40000 \ ms^{-2}$ **0**

(ii) $\Delta p = F \Delta t$

    equal to area a under the graph

    so, area of rectangle $= 2200 \ N \times 5 \times 10^{-3}$ s
                            $= 11 \ Ns$

    area of triangle $= \dfrac{1000 \ N \times (4 \times 10^{-3} \ s)}{2}$
                        $= 2 \ Ns$

    so, area under curve $= 11 \ Ns - 2 \ Ns$
                            $= 9 \ Ns \ Q.E.D.$ **2**

(c) (i) $F \Delta t = 9 \ Ns$
        $ma \Delta t = 9 \ Ns$ **0**

(ii) $F \Delta t = 9 \ Ns$ $p = \dfrac{Fs}{t}$

    so, $F = \dfrac{9}{5 \times 10^{-3}} = 1800 \ N$ $= \dfrac{1800 \times 0.70}{5 \times 10^{-3}}$

    so, power $= \dfrac{work}{time}$ $p = \dfrac{1260}{5 \times 10^{-3}} = 252000 \ W$ **2**

(d) Newton's 3rd law states that every action has an equal but opposite reaction. This means that when the gun is fired, the gun exerts a force on the bullet, making it move foward. The gun recoils because the bullet exerts an equal, but opposite force on the gun, meaning that it moves in the opposite direction to the bullet with the same force   **2**

N/A

This answer achieved 11/15

This suggestion that the time is short so that the acceleration is hard to measure accurately is not true.

The physics is poor here. The assumption is that the difference in force over the last 2.0 ms is the important effect. But it is the overall force that is acting, not a force difference.

The student divided the graph up into four rectangles (not shown here) and has demonstrated clearly the stages in the calculation for each rectangle.

The units are incorrect (N not N s), but as the units are given in the question there is no penalty.

This is a clear solution that carries through the calculation to five significant figures, with a clear rounding to a more appropriate number of figures at the end.

This is the more usual solution using the average rate of transfer of kinetic energy. The method of working is very clear.

All marks are awarded here as the link between force, acceleration and recoil is clear, together with the statement of the third law, and the implication that the force on the bullet is equal and opposite to the force on the gun.

(a) There is such a short time period (5 ms) (especially in comparison to the large change in acceleration) that the calculated value would be inaccurate and imprecise due to the likely uncertainties. **0**

(b) (i) 3 to 5 ms: $\Delta F = 200$ N

$F = ma$

$\Delta F = 2200$ N $- 2000$ N $= 200$ N $= 0.032$ kg $\times a$

6250 ms$^{-2} = a$

$= \dfrac{200 \text{ N}}{0.032 \text{ kg}}$ **0**

(ii) impulse $= F \times \Delta t =$ change in momentum

$=$ area under the graph $= A + B + C + D$

$= 0.5$ ms $(1100$ N$) + 0.5$ ms (1300)

$+ 2$ ms $(1750$ N$) + 2$ ms $(2100$ N$)$

$= 8900$ N ms $= 8.9$ N m ≈ 9 N **3**

(c) (i) Δ momentum $=$ impulse $= 9$ N s

$= m \Delta v = 0.032$ kg $\times \Delta v$

As initially at rest, $\Delta v = v = 281.25$ ms^{-1}

≈ 281 ms^{-1} **2**

(ii) power $= \dfrac{\text{energy transferred}}{\text{time taken}}$

$= \dfrac{\frac{1}{2} m v^2}{5 \text{ ms}} = \dfrac{\frac{1}{2} (0.032 \text{ kg}) (281 \text{ ms}^{-1})^2}{5 \text{ ms}}$

$= 252675$ Js$^{-1} \approx 253000$ W **3**

(d) Newton's third law states that every action will have an equal and opposite reaction. Therefore the bullet will exert a force on the gun equal to the force the gun exerted on it, and opposite in direction. The force on the gun gives an acceleration by Newton's second law which gives the gun speed backwards. Thus the gun will move, or recoil in that direction. **3**

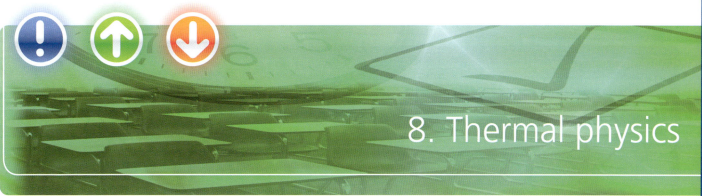

8. Thermal physics

Key terms for this chapter

- thermal equilibrium
- temperature—a measure of the average random kinetic energy per molecule
- internal energy of a substance and thermal energy (heat)
- mole—the amount of substance containing as many elementary particles as there are in 0.012 kg of carbon-12
- molar mass—the mass of one mole
- Avogadro constant—number of particles in one mole of a substance
- thermal (heat) capacity—change in thermal energy for 1 deg change in temperature
- specific heat capacity—thermal capacity per unit mass
- specific latent heat—thermal energy to change the phase of 1 kg of a substance
- pressure—force exerted on unit area
- equation of state of an ideal gas
- ideal and real gases—an ideal gas is one that obeys the equation of state for an ideal gas
- isochoric, isobaric, isothermal and adiabatic—terms that describe the process being applied to a gas

Thermal concepts

You should know:

- the relation between the Kelvin and Celsius temperature scales.

You should be able to:

- determine the direction of thermal energy transfer between objects
- define the mole, molar mass and the Avogadro constant
- distinguish between the concepts of temperature, internal energy and thermal energy (heat).

Example

A solid is at an initial temperature of 500 K. The solid is heated so that its temperature rises by 50 K. What are the initial temperature and the temperature rise of the solid, as measured on the Celsius scale of temperature?

	Initial temperature/°C	Temperature rise/deg
A.	227	50
B.	227	323
C.	773	50
D.	773	323

Answer: A

This question requires an understanding of the connection between Kelvin and Celsius temperature scales.

Kelvin scale: 0 K is also known as absolute zero and is effectively the temperature at which all molecular motion stops.

Celsius scale: 0°C is one of the fixed points of the Celsius scale and corresponds to the temperature at which water freezes to form ice (under standard conditions).

The relationship between the two scales is

$$T_{in\,K} = \theta_{in\,°C} + 273$$

Thermal concepts (continued)

The interval of one degree is the same "width" on both scales. The temperature difference between water freezing and boiling is 100 degrees on both scales. The correct way to write the difference between the temperatures of boiling and freezing water is 100 deg.

Be prepared

- Thermal energy flows from hot to cold regions, in other words, from a high temperature to a low temperature.
- Be ready to convert temperatures in both directions between the Kelvin and Celsius scales.
- You should learn the definitions of the mole, molar mass and the Avogadro constant. You can expect to be asked to use these quantities in calculations and to define them.
- The concepts of temperature and thermal energy (heat) should not be confused. One way to keep them separate

in your mind is to think of a spark that jumps from a log fire. The spark is at a high temperature but it may not burn you because it has a relatively small amount of thermal energy to transfer to your skin.

- Equally, the concepts of internal energy and thermal energy are different. Thermal energy is the energy transferred in a non-mechanical form. Objects do not possess a thermal energy—the object itself has a quantity of internal energy "stored" inside it.
- Internal energy is the sum of the total potential and the total random kinetic energy of the molecules of a substance.
  - Potential energy is due to the bond energy and the intermolecular forces of attraction between molecules.
  - The random kinetic energy is mainly due to the various motions of the molecules that make up the material.

Specific heat capacity, phase changes and latent heat

You should know:

- the differences between solids, liquids and gases in terms of the molecular structure and the motion of the particles.

You should be able to:

- define specific heat (thermal) capacity, thermal capacity and specific latent heat, and solve problems using them
- distinguish between evaporation and boiling
- explain in terms of molecular behaviour why temperature does not change during a phase change.

Example

A block of metal at a temperature of 90°C is placed in a beaker of water at a temperature of 0°C. The mass of the metal block and the mass of the water are equal. The final temperature of the water and the metal block is 9°C. Which of the following is the best estimate of the ratio

$$\frac{\text{specific heat of water}}{\text{specific heat of metal}}?$$

A. $\frac{1}{10}$

B. $\frac{1}{9}$

C. 9

D. 10

Answer: C

The equation that defines specific heat capacity is

specific heat capacity
$$= \frac{\text{energy supplied to the object}}{\text{mass of object} \times \text{change in temperature of object}}$$

The metal block is cooling down and all the thermal energy that it releases goes to heat up the water. You will need to assume that no energy is given to the surroundings. The masses of water and metal are equal, so they will cancel when you equate the energies.

Be prepared

- There are distinct differences between solids, liquids, and gases (phases of matter) in terms of:
  - the motion of the atoms or molecules in the phase
  - the spacing of the molecules in the phase
  - the density of the material in the phase.

  You should be aware of the similarities and differences between the phases.

- The word "specific" placed in front of the term for a physical quantity means "per unit mass". So, because the thermal capacity of an object is the amount of energy required to raise its temperature by 1 K, the specific

Specific heat capacity, phase changes and latent heat (continued)

thermal capacity of an object is the amount of energy required to raise 1 kg of the object by 1 K. Similarly, the specific latent heat is the energy required to change the state of 1 kg of a substance.

- The units of specific heat capacity are $J\,kg^{-1}\,K^{-1}$. The units of specific latent heat are $J\,kg^{-1}$.

- During **evaporation**, molecules in the liquid state transfer to the gaseous state. It occurs at temperatures below the boiling point and is purely a surface effect.

When **boiling** occurs, gas (in the form of bubbles) appears in the bulk of the liquid. Make sure that you can describe both effects, and the differences between evaporation and boiling.

- At a phase change (from liquid to gas, or solid to liquid), the energy supplied to the substance is breaking the intermolecular bonds between the molecules. During boiling, the molecules become free and become part of the gas that results from the process.

Kinetic model of an ideal gas

You should know:
- the assumptions of the kinetic model of an ideal gas
- that temperature is a measure of the average random kinetic energy of the molecules of an ideal gas.

You should be able to:
- define pressure
- explain how the molecular movement of an ideal gas results in its macroscopic behaviour.

Example

A sample of an ideal gas is contained in a cylinder. The volume of the gas is suddenly decreased. A student makes the following statements to explain the change in pressure of the gas.

 I. The average kinetic energy of the gas atoms increases.
 II. The atoms of the gas hit the walls of the cylinder more frequently.
 III. There are more atoms that are able to collide with the walls of the cylinder.

Which of these statements are true?

A. I and II only
B. I and III only
C. II and III only
D. I, II and III

Answer: A

Take special care when reading this question and questions like it. The gas is "contained in a cylinder". What does this imply

for the number of atoms? The volume of the gas is suddenly decreased. This allows you to predict the direction of any pressure and temperature changes. Pressure is related to the momentum change at the walls of the cylinder. Temperature is related to the average kinetic energy of the gas atoms.

Be prepared
- Pressure is defined as the normal force acting on a surface per unit area. It is measured in pascals (Pa), with $1\,Pa = 1\,N\,m^{-2}$.

- Learn the assumptions of the kinetic model. Help your memory by arranging them in a consistent way, beginning with the most basic (normally the statement that a gas consists of small particles), linking through the ideas of size to the statements about the forces acting on the particles.

- The temperature of a gas is a measure of the average random kinetic energy possessed by the gas molecules. Temperature measurements for liquids and solids follow from this.

- The great achievement of the kinetic model was that a theory could be constructed of how an ideal gas **ought** to behave. Most real gases behave approximately as ideal gases when the conditions are correct (low pressure and so on). So the experiments and the theory link.

- The kinetic model can model the way that molecules collide with the walls of a container. This is expressed in terms of momentum change per second and leads to a meaning of gas pressure. Ensure that you can explain this link in detail—if necessary, mathematically.

Gas laws

You should know:

- the equation of state for an ideal gas
- what is meant by absolute zero
- the concept of the Kelvin temperature scale.

You should be able to:

- describe the difference between an ideal gas and a real gas
- solve problems using the equation of state of an ideal gas.

Example

Two identical boxes X and Y each contain an ideal gas. In box X there are n moles of the gas at temperature T and pressure P_X. In box Y there are $3n$ moles of the gas at temperature $0.25T$ and pressure P_Y. What is the ratio $P_X : P_Y$?

A. $3:4$

B. $4:3$

C. $3:1$

D. $4:1$

Answer: B

This question requires that you know how to use the expression for the equation of state of an ideal gas:

$$\frac{PV}{T} = nR$$

P, V and T have their usual meanings, n is the number of moles of gas, and R is the universal gas constant. Substitute the quantities into the expression, cancelling symbols where necessary.

Be prepared

- Get plenty of practice in solving questions involving the equation of state for a gas. There are two forms of the equation. Units can be a particular problem here.

$$\frac{PV}{T} = nR$$

- P must be in Pa, V must be in m^3, n is in moles, and R is 8.31 J K^{-1} mol^{-1}

$$\frac{P_1V_1}{T_1} = \frac{P_2V_2}{T_2}$$

- P and V can be **any** correct pressure and volume units provided they are the **same** on both sides of the equation. However, T can **only** be in kelvin. If you cannot see why, imagine that the temperature in one of the equations is 273 K (0°C) and think about the effect of using Celsius in this case.

- A theoretical **ideal** gas cannot condense to form a liquid (or a solid). Everyday gases that do this are called **real** gases. However, these can behave as ideal gases for a limited range of conditions. Real gases have non-ideal behaviour at high pressures and low temperatures (when the molecules tend to be close together and have more influence on each other between collisions).

First law of thermodynamics

You should know:

- what is meant by the state change of a gas
- the first law of thermodynamics
- that the first law is a statement of energy conservation
- what is meant by a thermodynamic cycle.

You should be able to:

- deduce the expression for the work involved in a volume change of a gas at constant pressure
- describe changes of state for an ideal gas that are either isochoric (isovolumetric), isobaric, isothermal and adiabatic
- draw and annotate P–V diagrams
- calculate the work done in a thermodynamic cycle using its P–V diagram
- solve problems that involve the state change of a gas.

Example

The first law of thermodynamics may be expressed in terms of the quantities below:

ΔU, the **increase** in the internal energy of the system

Q, the energy transferred **to** the system by heating

W, the work done **on** the system

Which of the following is a correct statement of the law?

A. $W = \Delta U + Q$
B. $W = -\Delta U - Q$
C. $W = \Delta U - Q$
D. $W = -\Delta U + Q$

Answer: C
The first law of thermodynamics is an expression of the conservation of energy. The formulation that is expected in IB physics exams is

$$Q = \Delta U + W$$

but you may see this equation written in different ways in other textbooks. The equation applies to a thermodynamic system, which might be a gas in a cylinder, or a piece of hot metal cooling down in water, and so on. You need to be very clear about the meanings of the symbols and the positive or negative signs assigned to them, and this is the point of the question here.

+Q means that thermal energy is added to the system.

ΔU is the change in the internal energy of the system—if it is positive the internal energy increases in value.

+W means that work is done by the system on the surroundings.

To help you remember this, think of the equation as representing, on the left-hand side, the thermal energy being added to the system. The right-hand side has two terms: the energy appearing as internal energy, together with the work that the system does on its surroundings.

In answering this question, read the definitions of the symbols carefully.

Be prepared

- The meaning of the first law was discussed in the explanation of the multiple-choice question above. Make sure that you have a clear understanding of the symbols and their meaning and how the sign represents the direction of energy flow.

- A thermodynamic cycle is a series of processes in which the working substance (usually a gas) eventually comes back to its initial state.

- A gas that expands (changes its volume) at constant pressure P does an amount of work $P\Delta V$ where ΔV is the volume change. If the pressure is in Pa ($N\,m^{-2}$) and the volume change is in m^3, then the work done is $N\,m^{-2} \times m^3$, in other words, N m or J.

- This is also a good way to remember that the area under a P–V diagram is the work done by a gas in changing its state. If the gas is working in a cycle, then the work done in the cycle is the area enclosed by the lines on the P–V diagram.

Second law of thermodynamics and entropy

You should know:

- the second law of thermodynamics
- that thermal energy cannot spontaneously transfer from a low-temperature region to a high-temperature region
- that the entropy of a system expresses the degree of disorder in the system.

You should be able to:

- state the second law in terms of entropy changes
- discuss the entropy changes in examples of natural processes.

Example

A beaker contains water at a higher temperature than its surroundings. Which of the following correctly describes the entropy change in the water and in the surroundings as the water cools?

	Entropy of the water	Entropy of the surroundings
A.	decreases	decreases
B.	decreases	increases
C.	increases	decreases
D.	increases	increases

Answer: B

Entropy is a measure of the disorder in a system. The second law of thermodynamics predicts that the disorder in the universe increases as time goes on and as things happen. However, this does not mean to say that, in a particular and well-defined part of the universe, entropy can never decrease. The question here focuses on a system— the water in the beaker—that is part of a larger system. This larger system includes the beaker and its surroundings. The water cools and transfers some of its energy to its surroundings, which are, consequently, increasing in temperature. The water molecules are therefore moving slower and have less disorder. But the surroundings have more disorder.

Be prepared

- There are a number of ways to express the second law of thermodynamics. IB physics chooses one of the simplest forms: thermal energy cannot pass spontaneously from a region of low temperature to a region of high temperature. Disorder in the universe is described through the idea of entropy. Entropy is a quantity associated with a system and can be interpreted as the amount of disorder in the system.

- Be prepared to describe and discuss entropy changes in natural processes. You yourself are a good example of entropy in action. You are a highly ordered system, but you have only achieved this high order at the expense of disorder created elsewhere!

A2. This question is about thermal energy transfer.

(a) A piece of copper is held in a flame until it reaches thermal equilibrium. The time it takes to reach thermal equilibrium will depend on the thermal capacity of the piece of copper.

    (i) Define *thermal capacity*. *[1]*

    (ii) Outline what is meant by thermal equilibrium in this context. *[1]*

(b) The piece of copper is transferred quickly to a plastic cup containing water. The thermal capacity of the cup is negligible. The following data are available.

Mass of copper	= 0.12 kg
Mass of water	= 0.45 kg
Rise in temperature of water	= 30 K
Final temperature of copper	= 308 K
Specific heat capacity of copper	= 390 J kg K^{-1}
Specific heat capacity of water	= 4200 J kg K^{-1}

    (i) Use the data to calculate the temperature of the flame. *[3]*

    (ii) Explain whether the temperature of the flame is likely to be greater or less than your answer to (b)(i). *[2]*

[Taken from SL paper 2, November 2009]

How do I approach the question?

(a) (i) Recall the definition of thermal capacity (**not** specific thermal capacity).
(ii) Thermal equilibrium implies that the macroscopic parameters of a system have ceased to change with time. But the question asks for the answer to be in the context of the copper in the flame. Your answer must discuss this system.

(b) (i) This is an application of the defining equation for specific heat capacity. The thermal energy transferred from the copper is equal to the thermal energy gained by the water. Calculate the first and equate it to the second.
(ii) Although the copper is transferred "quickly", it is still likely that its temperature will drop during the transfer. Consider the effects on the transferred energy and therefore the accuracy of the estimate of the temperature of the flame.

Which areas of the syllabus is this question taken from?

- This question involves a knowledge of "Thermal concepts" (topic 3.1) and "Thermal properties of matter" (topic 3.2).

This answer achieved 1/7

The student is confusing melting/boiling point with thermal capacity.

The concept of thermal equilibrium here refers to a balance between flame and copper, not within the copper.

The words "in this context" in the question are crucial and mean that, to score, an answer must refer to the copper and flame.

The equation is mass × specific heat capacity × temperature = energy transfer. The manipulation of the equation is incorrect (and not shown).

(a) (i) The highest temperature a solid becomes before transitioning to a liquid/gas. **0**

(ii) In this context thermal equilibrium means that the copper, in it's entirety has reached the same temperature. Not a part of it is a different temperature. **0**

(b) (i) $390 \cdot .12 = 46.8$
$46.8 + 308 = 355\ °K$ **0**

(ii) The temperature of the flame is likely to be greater than a calculated estimate due to subtle variations in temperature based on different parts of flame and the presence of air/drafts. **1**

There is one mark awarded for the correct direction of change in the temperature.

This answer achieved 5/7

This is a good definition of **specific** thermal capacity and so is incorrect here.

This is a very basic statement and only just scores. The idea of thermal equilibrium is much more to do with energy balance rather than equality of temperature.

The student equates the energy lost by the copper to the energy gained by the water. However, the 308 K should be added to the 1211 K to arrive at the flame temperature. Here it is subtracted.

This is a good answer that focuses both on the change to the estimate and on the reason for the change.

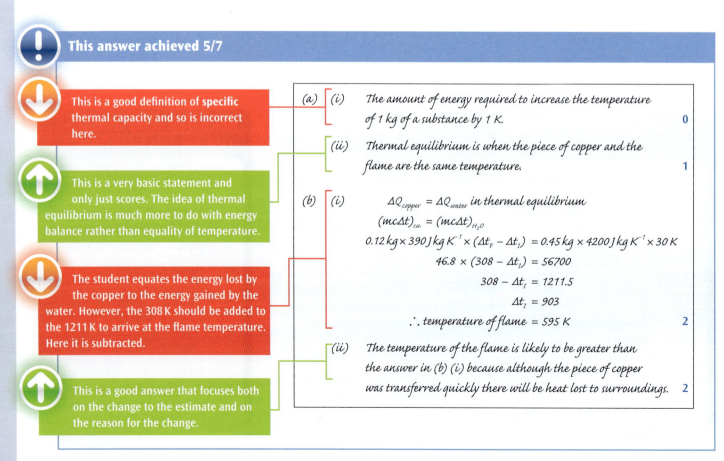

(a) (i) The amount of energy required to increase the temperature of 1 kg of a substance by 1 K. **0**

(ii) Thermal equilibrium is when the piece of copper and the flame are the same temperature. **1**

(b) (i) $\Delta Q_{copper} = \Delta Q_{water}$ in thermal equilibrium
$(mc\Delta t)_{cu} = (mc\Delta t)_{H_2O}$
$0.12\,kg \times 390\,J\,kg\,K^{-1} \times (\Delta t_F - \Delta t_I) = 0.45\,kg \times 4200\,J\,kg\,K^{-1} \times 30\,K$
$46.8 \times (308 - \Delta t_I) = 56700$
$308 - \Delta t_I = 1211.5$
$\Delta t_I = 903$
$\therefore$ temperature of flame $= 595\,K$ **2**

(ii) The temperature of the flame is likely to be greater than the answer in (b) (i) because although the piece of copper was transferred quickly there will be heat lost to surroundings. **2**

This answer achieved 7/7

This is a clear and accurate statement.

The student covers all possible areas in the answer: temperature equality and no net transfer of thermal energy.

The solution is clear and well presented. The student uses a sensible notation to remind the examiner of the distinction between temperatures and changes in temperature.

The student has not picked up the point that the thermal capacity of the cup is negligible, and therefore lists this as a reason for a low temperature estimate. There is no deduction for this, as the other reason is there too.

(a) (i) The amount of energy required to raise the temperature of a sample of material by 1 K. **1**

(ii) Thermal equilibrium occurs when the copper and flame are at the same temperature and no net transfer of thermal energy occurrs between them. **1**

(b) (i)
$$\Delta E_{copper} = -\Delta E_{water}$$
$$0.12 \times 390 \times \Delta T_{copper} = -(0.45 \times 4200 \times 30)$$
$$\Delta T_{copper} = -1212 \ K$$
$$T_{flame} = 308 + 1212$$
$$= 1520 \ K$$
$$= 1.5 \times 10^3 \ K$$
3

(ii) The temperature is likely to be greater because some thermal energy would have been lost to the surroundings while the copper was being moved or transferred to the cup while the copper was immersed. **2**

B3. This question is about internal energy, heat and ideal gases.

(a) The internal energy of a piece of copper is increased by heating.

Explain what is meant, in this context, by internal energy and heating. *[3]*

Internal energy:

Heating:

(b) An ideal gas is kept in a cylinder by a piston that is free to move. The gas is heated such that its internal energy increases and the pressure remains constant. Use the molecular model of ideal gases to explain

(i) the increase in internal energy. *[1]*

(ii) how the pressure remains constant. *[3]*

[Taken from SL paper 2, time zone 1, May 2009]

How do I approach the question?

(a) This question requires that you give a clear distinction between internal energy and the thermal energy that is transferred when the copper is heated.

(b) The question does not use the term "kinetic model of an ideal gas". But this theory should be in your mind as you answer the question. Explain in terms of the behaviour of the gas molecules why (i) the internal energy increases and (ii) the pressure remains constant.

Which areas of the syllabus is this question taken from?

- This question requires knowledge of "Thermal concepts" (topic 3.1) and "Thermal properties of matter" (topic 3.2).

This answer achieved 0/7

This answer seems to focus on the stored chemical potential energy as the student writes about a (chemical) reaction.

(a) Internal energy: the amount of energy given off by a the piece of copper before reacted with anything

Again, there is a consideration of a chemical process, which is not what is happening when the copper is heated.

Heating: the chemical process of expanding the intermolecular forces of copper   **0**

While it is true that the gas molecules move farther apart, this does not account for the change in internal energy.

(b) (i)   There is an increase in internal energy because the particles for the gas move far apart.   **0**

(ii)   The pressure remains constant because the internal energy remained constant and because the area of space was not effected.   **0**

This contradicts (b)(i) as it suggests that internal energy is now constant.

✓ Contradictory statements are unlikely to gain marks. They imply that the student does not understand the work in either part.

This answer achieved 5/7

This is a clear statement about the microscopic meaning of internal energy.

(a) Internal energy: the sum of the kinetic energy of the particles   **2**

Heating: process by which the average kinetic energy of the particles is increased   **0**

There are ways to increase the average kinetic energy of an object other than by heating. The answer is not specific enough.

(b) (i)   The molecules are gaining kinetic energy   **1**

(ii)   Because of the increase in kinetic energy, the molecules move faster and the rate at which they collide with the piston is increased. This creates a greater force on the piston, which moves back from the gas, increasing the volume until the rate of collisions balances out with the increased speed.   **2**

Here the idea of gaining kinetic energy is enough to link to the increase in internal energy. However, it would be better to discuss either the total kinetic energy or the average kinetic energy.

Two good points are made: increase in kinetic energy means that the molecules are moving faster; so the rate of collision with the piston and hence the force increases. However, a third point is omitted.

This answer achieved 6/7

The phrase "amount of kinetic energy in molecules" could be better expressed, but the idea is there.

Here there is a clear mention of the transfer of energy that leads to the increase in kinetic energy.

Try to be careful with your use of English. What does "vibrate faster" mean? Perhaps "vibrate with a higher frequency" or "move faster" would have been better.

Exam papers are designed so that there is plenty of room for your answer. This student is cramming a lot of unnecessary information in here and going well over length. Carried to excess, this can mean that a student may not finish the entire paper.

Again, a clear link is made between an increase in internal energy and an increase in the kinetic energy (hence, speed) of the molecules.

This is an answer that links kinetic energy and speed, and points out that movement of the piston increases the volume of the gas. Therefore there is more room for the molecules to move in, and more time between collisions. So, even though there is an increased speed, the force per unit area (pressure) remains constant.

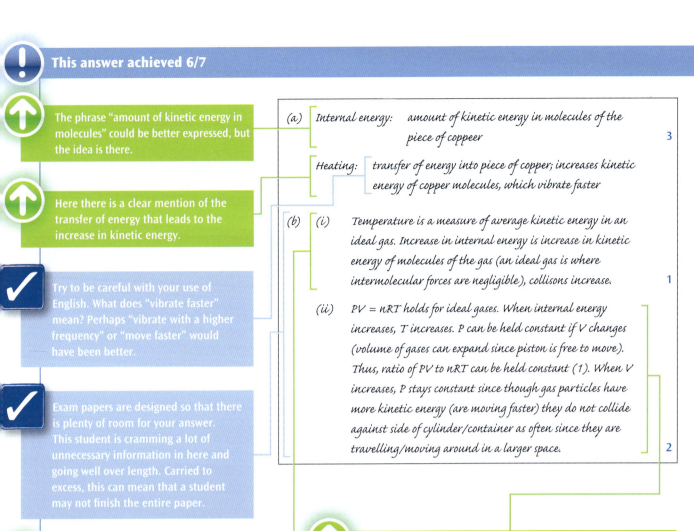

(a) Internal energy: amount of kinetic energy in molecules of the piece of coppeer   3

Heating: transfer of energy into piece of copper; increases kinetic energy of copper molecules, which vibrate faster

(b) (i)   Temperature is a measure of average kinetic energy in an ideal gas. Increase in internal energy is increase in kinetic energy of molecules of the gas (an ideal gas is where intermolecular forces are negligible), collisons increase.   1

(ii)   PV = nRT holds for ideal gases. When internal energy increases, T increases. P can be held constant if V changes (volume of gases can expand since piston is free to move). Thus, ratio of PV to nRT can be held constant (1). When V increases, P stays constant since though gas particles have more kinetic energy (are moving faster) they do not collide against side of cylinder/container as often since they are travelling/moving around in a larger space.   2

9. Oscillations, waves and wave phenomena

Key terms for this chapter

- displacement x, amplitude x_0, frequency f, time period T, wavelength λ, wave speed c, intensity, phase difference ϕ and angular frequency ω
- kinetic energy, potential energy and total energy during simple harmonic motion (SHM)
- damping, natural frequency, forced frequency and resonance
- progressive (travelling) wave—wave that propagates energy
- standing (stationary) wave—wave for which there is no energy propagation between particles of the wave
- transverse wave—direction of energy propagation is at 90° to motion of particles of medium
- longitudinal wave—direction of energy propagation is parallel to direction of particle motion
- wavefront and ray
- crest and trough of a wave
- compression and rarefaction for longitudinal waves

Kinematics of simple harmonic motion (SHM)

You should know:

- the defining equation of SHM
- definitions for the terms used in SHM (all given in the key terms).

You should be able to:

- apply the equations

$$v = v_0 \sin \omega t$$
$$v = v_0 \cos \omega t$$
$$v = \pm \omega \sqrt{(x_0^2 - x^2)}$$
$$x = x_0 \cos \omega t$$
$$x = x_0 \sin \omega t$$

- solve SHM problems using graphs and calculations.

Example

Which graph correctly shows how the acceleration a of a particle undergoing SHM varies with its displacement x from its equilibrium position?

A.
B.

C.
D.

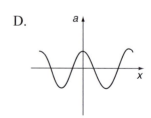

Answer: A

The defining equation for simple harmonic motion is $a = -\omega^2 x$, where a is the acceleration, x is the displacement and ω is a constant. Note the negative sign. So, in SHM, a is proportional to −x.

Kinematics of simple harmonic motion (SHM) (continued)

Be prepared

- An object is said to be moving with simple harmonic motion if:
  - the object's acceleration is directly proportional to its displacement, and
  - the vector direction of the acceleration is opposite to that of the displacement.
- Ensure that you can define the important key terms in SHM.
- The defining equation leads to an oscillating motion. The displacement–time graph is a sine or cosine graph (or one that begins somewhere between the zero and extreme positions).

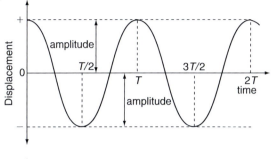

- The speed v of an object undergoing SHM is, as usual, the rate of change of the displacement. Similarly, the acceleration is the rate of change of velocity.

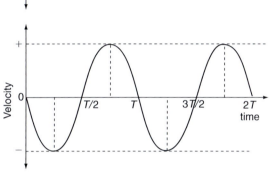

- Ensure that you can sketch the graphs of a–x, v–x, a–t and x–t for a particular SHM if one of them is given to you. The key idea is that the a–t graph is the x–t graph inverted (upside down), while the v–t graph is out of phase by 90°.

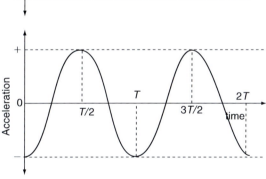

Energy changes during SHM

You should know:

- how the kinetic energy and potential energy interchange during SHM.

You should be able to:

- solve energy-change problems using the equations

  $E_K = \frac{1}{2}m\omega^2(x_0^2 - x^2)$ for the kinetic energy of a particle undergoing SHM

  $E_T = \frac{1}{2}m\omega^2x_0^2$ for the total energy

  $E_P = \frac{1}{2}m\omega^2x^2$ for the potential energy.

Example

A mass on the end of a horizontal spring is displaced from its equilibrium position by a distance A and released. Its subsequent oscillations have total energy E and time period T.

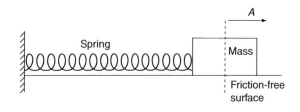

Energy changes during SHM (continued)

An identical mass is attached to an identical spring. The maximum displacement is $2A$. Assuming this spring obeys Hooke's law, which of the following gives the correct time period and total energy?

	New time period	New energy
A.	T	$4E$
B.	T	$2E$
C.	$\sqrt{2}T$	$4E$
D.	$\sqrt{2}T$	$2E$

Answer: A

The time period of the spring is related to the angular frequency ω of the system. For a spring system, ω depends on the spring constant of the spring and the mass that is oscillating. The question makes it clear that neither of these quantities change. Look carefully at the equations for SHM energy at the beginning of this section. The total energy is always related to x_0^2 (amplitude^2).

Be prepared

- In SHM the total energy constantly transfers between kinetic and potential—entirely kinetic when the system is moving at its highest speed (at the equilibrium point), entirely potential when the object is stationary (at the extreme points of the motion). In between these two points, the total energy is a mixture of kinetic and potential.

- Be able to sketch graphs of kinetic, potential and total energy against time and also against displacement. These energy–time graphs have a cycle time half that of the SHM itself. The graphs are never negative, always staying above the $E = 0$ axis. The shape is $\sin^2$, which is similar to, but not quite the same as, a sine curve. The energy–displacement graphs are parabolas. The total energy for true SHM is always constant.

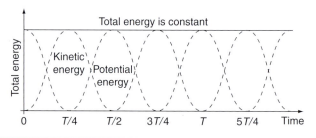

Forced oscillations and resonance

You should know:

- what is meant by the natural frequency of vibration, forced oscillations and resonance

- what is meant by over- and under-damping, and by critical damping.

You should be able to:

- give examples of damped oscillations

- describe, using a graph, how the amplitude of an object varies with the forced frequency of the vibration.

Example

The graph below represents the variation with time of the displacement of an oscillating particle.

The motion of the particle is

A. over-damped.

B. critically damped.

C. lightly damped.

D. not damped.

Answer: C

As mentioned earlier, true SHM has a constant amplitude and a constant total energy. Real life is different because there are damping losses in the system (such as air resistance or friction) that reduce the energy and hence the amplitude as time goes on. All four types are listed in the responses to this question. Heavy damping reduces the amplitude quickly, typically in a time less than one period. Critical damping is a special case where the damping reduces the movement of the system to zero in the shortest possible time.

Be prepared

- An oscillating system, if displaced and released, will oscillate with a **natural frequency** of oscillation that

Forced oscillations and resonance (continued)

is determined by the system. For a mass on a spring, the mass and the spring constant set the natural frequency.

- An oscillating system can be forced to oscillate by applying a periodic force to it. Such an oscillation is said to be **forced**. If the two frequencies, natural and forcing, are closely matched, then the system can resonate, and the system will oscillate with a large amplitude, possibly large enough to destroy the system.

- Examples of driven systems include a car suspension and a child on a swing pushed repeatedly.

- Ensure that you can sketch graphs of displacement against time for the cases of over-, under- and light damping. You should also be prepared to describe these three degrees of damping and also to give examples of each type of motion.

- Check that you can draw the graph that shows how the amplitude of an object varies with the forcing frequency

of the vibration. Examples of the graph can be found in standard textbooks. Your graph should bring out several features.

- As damping increases, the maximum amplitude at resonance decreases.

- The resonant frequency does not equal the natural frequency except when the damping is zero.

- The resonant frequency drifts to lower forcing frequencies as the damping increases.

- The graph is not symmetrical about the natural frequency.

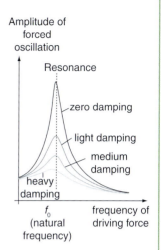

Wave characteristics

You should know:

- the definitions of displacement, amplitude, frequency, period, wavelength, wave speed and wave intensity

- that all electromagnetic waves travel at the same speed in free space

- what is meant, for waves, by crest, trough, compression and rarefaction.

You should be able to:

- describe a wave pulse and a travelling wave

- give examples of transverse and longitudinal waves and describe them

- describe wavefronts and rays

- draw and understand displacement–time and displacement–position graphs for transverse and longitudinal waves

- use the equation wave speed = wavelength × frequency

- recall the orders of magnitudes of wavelengths of the important radiations in the electromagnetic spectrum.

Example

In which of the following regions of the electromagnetic spectrum is radiation of wavelength 600 nm located?

A. microwaves

B. radio waves

C. visible light

D. X-rays

Answer: C

You are expected to recall the approximate wavelengths for the important regions of the electromagnetic spectrum. In addition to the four regions above, you should be aware of the wavelengths for infrared and ultraviolet, and gamma rays.

Be prepared

- Many of the terms used in wave theory are similar to those that occur in SHM. Additional ideas include wavelength, wave speed and wave intensity.

- All types of waves have "troughs" and "crests" that correspond to the maxima and minima of the

Wave characteristics (continued)

displacement. Electromagnetic waves can be imagined in these terms if the crests and troughs relate to the graph of amplitude of electric/magnetic fields against time.

- Sound waves have compressions and rarefactions. These do not occur at the maximum and minimum displacement points but at the points where displacement is zero. You should make sure that you understand how the compressions and rarefactions arise for a sound wave and how they relate to the wave shape.

- Electromagnetic waves do not need a material (medium) through which to travel. They have some properties that distinguish them from waves associated with media. All electromagnetic waves travel at 3.0×10^8 m s^{-1} in a vacuum. They slow down, usually by different amounts depending on wavelength, when they enter a medium.

- The energy of transverse waves (electromagnetic waves, waves on a string) moves in a direction at 90° to the oscillation direction of the particles of the medium or the electric/magnetic fields. For longitudinal waves (sound waves) the energy direction and the direction of oscillation of the medium are the same. You should be able to draw displacement–time and displacement–position graphs for both types of waves.

- The relationship between wavefronts and rays is straightforward. A ray gives the direction of the movement of a wave at a point in space. The ray is at 90° to the wavefront at this point.

- In solving problems using the wave equation, take care that the units match. If the speed is in m s^{-1}, the wavelength must be in m with the frequency in Hz.

Wave properties

You should know:

- Snell's law
- what is meant by diffraction at apertures and obstacles
- what is meant by superposition.

You should be able to:

- describe the reflection and transmission of waves when they meet a boundary between two media
- explain what is meant by constructive and destructive interference and apply conditions for these two effects in terms of both path and phase difference
- determine the resultant of two waves using the principle of superposition.

Example

Plane wavefronts are incident on a boundary between two media labelled 1 and 2 in the diagram. The diagram of the wavefronts is drawn to scale.

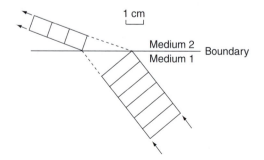

Which of the following is the ratio of the refractive index of medium 2 to that of medium 1?

A. 0.50

B. 0.67

C. 1.5

D. 2.0

Answer: B

The ratio of the refractive indices is the same as the ratio of the wave speeds. The wavefronts are a wavelength apart. You can use a ruler to determine the ratio of the wavelengths. The wave equation indicates that the ratio of the wave speeds is equal to the ratio of the wavelengths (the frequency of the wave will be the same in both media).

Be prepared

- When a wave meets a boundary between two media, reflection back into the original medium and transmission into the new medium occur. Sometimes the reflection will be inverted (out of phase) relative to the original wave before reflection; this depends on the relative wave speeds. Snell's law relates the refractive index of a pair of media to the ratio of the wave speeds in the two media. You can expect questions that involve calculations using refractive index.

- When a wave is incident on an aperture or an obstacle, diffraction occurs. The wave spreads out and does not move as might be expected by a ray model, where the

Wave properties (continued)

rays travel in straight lines. The amount of diffraction observed depends on the dimension of the aperture or obstacle relative to the wavelength of the wave.

- Superposition is another effect that occurs for waves of all types. The key is to add the instantaneous displacements of the waves, taking into account the sign of the displacement. You may need to do this in several places along the line of a pair of waves in order to obtain a profile for the new superposed waveform.

- Superposition has an important part to play in constructive and destructive interference. If two waves arrive at the same point **in phase**, then they will add together to have a wave with an amplitude equal to the sum of the amplitudes. If two waves arrive at the same point 180° (or π radians) **out of phase**, then—if the waves have the same amplitude—the waves will exactly cancel. Another way to describe the phase difference is in terms of path difference.

Standing (stationary) waves

You should know:

- the nature of standing waves
- how one-dimensional standing waves are formed
- the modes of vibration in strings and open and closed pipes.

You should be able to:

- compare standing and travelling waves
- solve problems involving standing waves.

Example

The wavelength of a standing (stationary) wave is equal to

A. the distance between adjacent nodes.

B. twice the distance between adjacent nodes.

C. half the distance between adjacent nodes.

D. the distance between a node and an adjacent antinode.

Answer: B

The node on a standing wave is a point at which there is no displacement (the point of maximum displacement is an antinode). The two waves that superpose completely cancel at this point. Adjacent nodes correspond to points that must be separated by half a wavelength on the original travelling waves.

Be prepared

- Standing waves occur when two travelling waves moving in opposite directions superpose. The waves do not have to have the same amplitude, but they do need to have the same wavelength/frequency.

- You should be able to describe the similarities and differences between standing waves and travelling waves. There are a number of points to consider.
  - Energy is propagated in a travelling wave.
  - Energy is not propagated in a standing wave in the case where the travelling waves have equal amplitudes.
  - All possible phase differences occur in a travelling wave.
  - Only phase differences of 0, π, 2π and so on can occur in standing waves.
  - A travelling wave has a single value of amplitude.
  - All amplitude values between 0 and a maximum occur in a standing wave.

- You should be able to sketch the first few harmonics present in a sounding pipe that is open at both ends (called simply "open") and in one that is closed at one end (called "closed").

- You should be able to sketch the first few harmonics present in a stretched string fixed at both ends.

- You should be able to explain how the standing waves form in strings and pipes.
  - Strings are straightforward. The wave reflects inverted at the fixed ends.
  - Pipes are more tricky. The fixed end acts as a barrier so that the air molecules cannot move longitudinally along the pipe at this end. A displacement node therefore forms. At open ends, the air pressure has to be equal to average air pressure at all times, so a pressure node (a displacement antinode) forms.

- Remember that, in both strings and pipes, one "loop" in the wave pattern corresponds to half a wavelength of the sounded frequency.

Doppler effect

You should know:

- what is meant by the Doppler effect.

You should be able to:

- explain the Doppler effect in terms of wavefront diagrams for cases when the detector or the observer are stationary
- apply the Doppler effect equations for sound.

Example

During a journey an observer travels at constant speed towards, and then goes beyond, a stationary emitter of sound.

Path followed by observer

• Stationary emitter of sound

The frequency of the sound as measured at the emitter is f. The frequency according to the observer

A. is always greater than f.
B. is always equal to f.
C. is always less than f.
D. varies during the journey.

Answer: D

When the observer is moving towards the sound waves, the observer crosses the emitted wavefronts more often than the source emits them. What does this mean for the observed frequency? As the observer moves away, wavefronts are crossed less often. What conclusion do you draw about the observed frequency for the whole of the event?

Be prepared

- The Doppler effect occurs when there is relative movement between a source of a wave and the observer of the wave. The effect is observed with all wave types. You will not be asked about the case where source and observer both move at the same time.

- Be ready to draw or interpret diagrams that show successive positions of the wavefronts. Practise drawing these as part of your revision. The Doppler effect is much easier to describe with a diagram than in words, providing you have the knack of drawing good circles of changing radius as the centres of these circles move along a straight line! Try it!

- Doppler calculations are often straightforward, but make sure that you are using the correct equation and the correct sign in the equation. This depends on whether the source/observer is approaching or receding.

Diffraction

You should know:

- how the relative intensity of light diffracted at a single slit varies with the angle of diffraction.

You should be able to:

- sketch and use the graph of variation with angle of diffraction of the relative intensity of light diffracted at a single slit
- derive the formula

$$\theta = \frac{\lambda}{b}$$

for the first minimum of the diffraction pattern, and use it to solve single-slit diffraction problems.

Example

Which diagram best shows diffraction of plane wavefronts at a single slit?

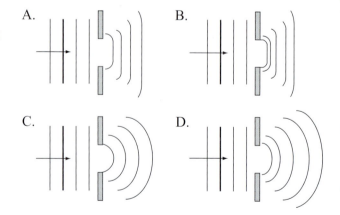

A.   B.

C.   D.

Answer: A

Questions such as this need care. The key point is to look carefully at the relationship between the aperture size and the wavelength. Diffraction is most obvious when these are of similar size. That is not the case here, the aperture is larger. Another

Diffraction (continued)

distracter here is the behaviour of the waves near the aperture. If there is no medium change, diffraction effects do not lead to a change in wavelength. Look carefully at the diagrams to see if a temporary wavelength change is implied.

Be prepared

- Drawing the graph of intensity against diffracting angle needs care. You need to show (see following graph):
  - relative heights of the maxima of intensity (most students draw the first maximum far too high)
  - relative widths of the angular spreads of the maxima (the central maximum is roughly double the spacings between other maxima)
  - a good representation of the shape of the curve.

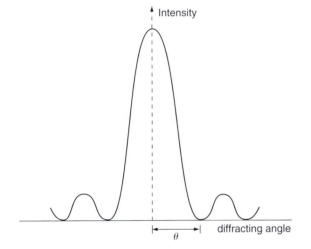

- In the equation

$$\theta = \frac{\lambda}{b}$$

θ tells you the angular position of the first minimum of the diffraction curve. At this point in space, the photons from the slit arrive so that their total effect is zero intensity. You will not be asked to consider other minima.

- In this equation above, θ is measured in **radians** not degrees. Remember that $2\pi\,\text{rad} = 360°$, and convert if necessary. Make sure that your calculator is set correctly too.

- Make sure that you can derive the equation. There are a number of ways to carry out this derivation that are shown in textbooks. Try to learn the basic physics of the derivation rather than the individual equations in the proof.

- The equation in the form

$$\theta = \frac{\lambda}{b}$$

is valid for a rectangular, straight-sided slit. If the aperture is in the form of a circle, the equation becomes

$$\theta = \frac{1.22\lambda}{b}$$

Marks can be lost for ignoring the 1.22 factor.

- Be ready for questions that ask for the effects of change on the diffraction pattern, but without expecting a full calculation. When you revise this topic, test yourself by asking what happens to the diffraction pattern if the slit becomes wider/narrower, if the light changes from red to blue, and so on. Be ready to draw the changed diffraction pattern in these cases.

Resolution

You should know:

- the Rayleigh criterion for the images of two sources to be just resolved.

You should be able to:

- describe the significance of resolution for the use of CDs, DVDs, the electron microscope and radio telescopes

- solve problems involving resolution

- sketch how the relative intensity of light emitted by two point sources diffracted by a single slit varies with angle of diffraction.

Example

Two stars close to each other as viewed from Earth both emit electromagnetic waves of wavelength $3.0 \times 10^{-2}\,\text{m}$. The waves are received by a radio telescope that has a collecting dish of diameter $60\,\text{m}$. The two stars are just resolved if their minimum angular separation in radians is of the order of

A. 2×10^{4}

B. 2×10^{2}

C. 6×10^{-2}

D. 6×10^{-4}

Resolution (continued)

Answer: D

The equation for the minimum angular separation θ is

$$\theta = \frac{1.22\lambda}{d}$$

where λ is the wavelength of the radiation and d is the diameter of the aperture that receives the radiation. One of the multiple-choice responses is meant to be the nearest estimate to the correct answer. Take care not to invert the expression and thus form the reciprocal of the answer. Reciprocals are often given as one or more of the responses in many multiple-choice questions.

Be prepared

- Resolution has great importance in many areas of optics and imaging. The list given in the syllabus is not exhaustive. The detail that can be seen in images formed by all telescopes and microscopes (optical and electron) is limited by the wavelength in use.

- When two adjacent sources are viewed through an aperture, diffraction occurs separately for both images, and the patterns are then combined by the detecting device. For the sources to be just resolved, the central maximum of one pattern must lie on the first minimum of the other. In other words, there must be a detectable drop in the total intensity pattern. If the patterns are farther apart than this, then they are resolved. If they are closer, then the patterns cannot be distinguished.

- Practise drawing the three diffraction patterns for the cases of: not resolved, just resolved and well resolved.

- Pay particular attention to the instruments mentioned in the syllabus.
    - Electron microscope: the wavelength of the electrons depends on the accelerating potential in use—the larger the potential difference, the larger the energy and the shorter the wavelength, so the better the resolution.
    - Radio telescopes: the larger the dish diameter, the smaller the angle θ (wavelength will be constant here), and so the better the resolution.

Polarization

You should know:

- what is meant by polarization and polarized light

- what is meant by:
    - polarizer and analyser
    - optically active substance

- how polarization is used in stress analysis

- how liquid-crystal displays work.

You should be able to:

- solve problems involving the polarization of light including the use of:
    - Brewster's law
    - Malus' law

- describe how polarization can be used to determine the concentration of some solutions

- describe how reflection leads to polarization.

Example

Unpolarized light is shone through two identical polarizers whose axes are parallel.

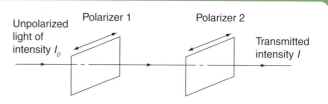

The ratio $\dfrac{I}{I_0}$ is

A. 100%.

B. 50%.

C. 25%.

D. 0%.

Answer: B

There are two ways to approach this problem. One is via Malus' law, which relates intensities before and after a polarizer. This equation is in the Physics data booklet. Another approach is via amplitudes, which are related by cos θ. Taking this route, you need to remember that intensity is (amplitude)².

Be prepared

- Longitudinal waves cannot be polarized; only transverse waves show polarization.

Polarization (continued)

- Electromagnetic radiation from many sources, including the Sun, is unpolarized. Sunlight arriving at Earth has electric field vectors at all possible angles. Polarizers can block unpolarized radiation so that only one direction of electric field vector remains. (The same applies to the magnetic field vector too, but this is not usually considered.)

- Make sure that you can describe how a polarizer works (think of a thin slit with an oscillating rope going through it) and that you understand how two polarizers together can be used to analyse the behaviour of an object between them. This idea leads to stress analysis in buildings, bridges and so on, where stress in the object leads to rotation of light, which can be detected with an analyser.

- Liquid-crystal displays (LCD) rely on polarization being switched on or off electronically in a thin material so that light travelling through the material can be blocked or allowed through on its way out.

- Brewster's law predicts the unique reflection angle at which unpolarized light can be completely plane polarized. At this angle, when the light is incident on a horizontal surface, the reflected light is all polarized in the plane of this surface. The oscillating electrons in the surface that are causing the reflection cannot move up and down out of the surface, only along the surface of the medium. Make sure that you understand the geometry of the situation, including which is the Brewster angle itself.

- You may be asked how solution concentrations can be measured using polarization. Make sure you can draw a clear labelled sketch of a possible arrangement and write a systematic account of the measurements needed to establish the result.

Part 2   This question is about the diffraction of light.

(a)   (i)   Describe what is meant by the diffraction of light.   *[2]*

   (ii)   A parallel beam of monochromatic light from a laser is incident on a narrow slit. The diffracted light emerging from the slit is incident on a screen.

The centre of the diffraction pattern produced on the screen is at C. On the axes sketch a graph to show how the intensity *I* of the light on the screen varies with the distance *d* from C.   *[3]*

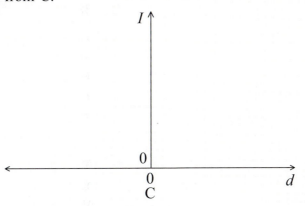

(iii) The slit width is 0.40 mm and it is 1.9 m from the screen. The wavelength of the light is 620 nm. Determine the width of the central maximum on the screen. *[3]*

(b) (i) When two separate lasers are used as sources, the images of the slit formed by the light from each laser are resolved. State what is meant by the term *resolved* in this context. *[1]*

(ii) A car, with its two headlights switched on, is approaching an observer who has good eyesight. Outline why, at a long distance from the observer, the images of the headlights of the car are not resolved by the observer. *[4]*

[Taken from HL paper 2, time zone 2, May 2009]

How do I approach the question?

(a) (i) Describe what happens when light is diffracted. There are two points to make, as shown by the mark allocation in the margin. Focus on the effect that occurs and the difference that happens because of the presence of an aperture or obstacle.

(ii) Sketch an intensity–distance graph for the diffraction pattern. Treat the distance axis as equivalent to the angle of diffraction mentioned in the syllabus. If you are attempting this question without looking at the student examples first, re-read the section on the points required on this sketch.

(iii) The equation is quoted in the *Physics data booklet*. Pay attention to the units in the question. You will need to evaluate the linear distance having calculated the angular spread.

(b) (i) As this is a "**state**" question, you do not need to explain the Rayleigh criterion, simply to make clear what the term "resolution" means.

(ii) This is the point at which a more detailed explanation is required. As this is an "**outline**", a full Rayleigh criterion treatment is still not required, but you should make clear the effect of diffraction pattern overlap on resolution.

Which areas of the syllabus is this question taken from?

- This question requires a knowledge of "Wave properties" (topic 4.5), "Diffraction" (topic 11.3) and "Resolution" (topic 11.4).

This answer achieved 2/13

 This is not clear enough. "Bending of light" is the way a student might describe refraction, and there is not enough here to describe diffraction at IB level.

There is 1 mark for the recognition that there is a central maximum but, as there are no clear minima (the graph does not extend far enough), there is no further credit.

The idea that the images can be seen separately is all that is required for the mark here as it is a "**state**" question.

This gains no credit. The idea of scattering is incorrect and the final equation is related to the Doppler effect, which is of no relevance in this question.

(a) (i) Light can bend around objects or holes.　　0

(ii)
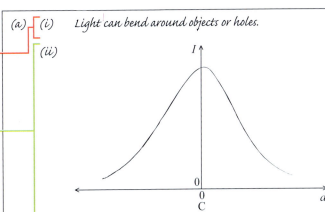

1

(b) (i) The images of the slit are separated so that they can be seen separately and distinguished from one another.

1

(ii) The light from the headlights scatters so that the light rays emitted appear to be coming from a single source. This is since

$$f' = f\left(\frac{v}{v \pm u_s}\right)$$

0

This answer achieved 6/13

 There are too many confused ideas here. The idea of separation is more like the dispersion that happens in a spectrum. Maxima and minima could be appropriate, but they are discussed in terms of interference. This is an effect associated with two or more overlapping regions of light. That is not what happens in diffraction.

A complete sketch with all required details is given (the subsidiary maxima are a little too high, but this is well within limits).

Notice how poor the sketch is on the negative *d* side of the graph. The trick here for right-handed people is always to draw from left to right—starting at large negative *d* and moving towards the origin, having marked the minima positions first.

These minima do not quite sit on the axis—and they should. The examiner has given benefit of the doubt, but this must have been a close thing.

 The solution begins well with the correct equation, but the substitution is incomplete. The conversion to metres is correct and the mention of the approximate equality of radian and degree is good.

 The idea that the images can be seen as separate is correct and clear.

The mention that ideas of resolution hinge on diffraction theory gains 1 mark.

(a) (i) Diffraction is the separation of light into maxima and minima (caused by constructive and destructive interference) **0**

(ii)   **3**

(iii) The central maximum is between first minima $\left(\pm \frac{\lambda}{b}\right)$

$b = 4.0 \cdot 10^{-4} m$ $\lambda = 6.20 \cdot 10^{-9}$

small angle, so $\theta \approx \sin\theta$

$\theta = \dfrac{0.0004\ m}{1.9\ m} =$ **1**

(b) (i) Resolved = the (maxima of the) light from the two lasers can be seen as separate lights   **1**

(ii) Eye of observer Car

pupil = b

The situation can be looked at so that the pupil is the slit for diffraction (b). As the distance between the eye and the car's headlights is very long, the angle between the two separate headlights is very small   **1**

There are no references to the overlap of diffraction patterns.

This answer achieved 11/13

There is the clear idea that diffraction is the spreading of light.

What is missing is the idea that the light spreading is beyond that predicted by the geometrical shape of the aperture or obstacle.

A good and complete sketch of the pattern. This example makes it very clear that the minima are at an intensity of zero.

This solution begins well. However, the student thinks that the answer is a distance; it is, in fact, an angle measured in radians. So, the final step to multiply the angle by the distance to the screen is missing.

This answer goes beyond what is required. It is almost describing the Rayleigh criterion. The idea of separate patterns is very clear.

The question asks for an "outline". This answer would not score so highly as an "explain". It is clear that diffraction is occurring at the eye of the observer with one diffraction pattern from each light. There is description of the patterns overlapping.

The physics is actually wrong because the Rayleigh criterion says that, for the images to be just resolved, the central maximum of one pattern sits over the first minimum of the other. As this was not required, the answer gains full marks.

(a) (i) When light is diffracted, its wavefronts spread to all directions around an object (or through a slit). 1

(ii)

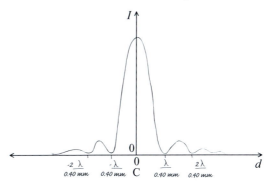

3

(iii) 1st mimimum occurs at $\dfrac{\lambda}{b} = \dfrac{620 \cdot 10^{-9}\ m}{0.40 \cdot 10^{-3}\ m} = 0.00155\ m$

Width of central maximum $= \dfrac{2\lambda}{b} = 0.0031\ m$

$= 3.1\ mm$ 2

(b) (i) Two separate central maxima are observed on the screen, on either side of C, and can be identified as the two lasers. 1

(ii)

..... diffraction pattern produced by 1st headlight

_ _ diffraction pattern produced by 2nd headlight

— combined diffraction pattern

For the two sources to be just resolved, the 1st minimum of one pattern should lie on top of the 1st maximum of the other pattern. At a long distance from the observer, this does not occur, but the two patterns overlap as to produce a combined pattern with only one central maximum. The diffraction occurs around the pupil of the observer. 4

B2. This question is about simple harmonic motion and waves.

An object is vibrating in air. The variation with displacement x of the acceleration a of the object is shown below.

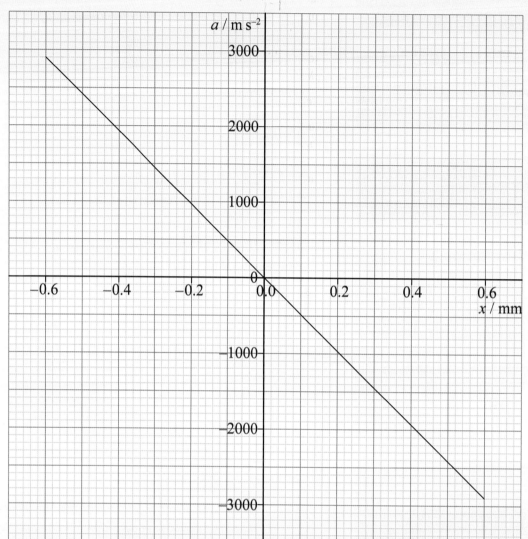

(a) State and explain **two** reasons why the graph opposite indicates that the object is executing simple harmonic motion. [4]

1.

2.

(b) Use data from the graph to show that the frequency of oscillation is 350 Hz. [4]

(c) State the amplitude of the vibrations. [1]

(d) The motion of the object gives rise to a longitudinal progressive (travelling) sound wave.

(i) State what is meant by a longitudinal progressive wave. [2]

(ii) The speed of the wave is 330 m s^{-1}. Using the answer in (b), calculate the wavelength of the wave. [2]

[Taken from SL paper 2, time zone 2, May 2009]

How do I approach the question?

(a)   This question asks for the interpretation of an acceleration–displacement graph. Consider the extent to which the defining equation is satisfied by the graph.

(b)   The defining equation is again the route to take. You can read off values for x_0 and x. Use these to calculate ω and hence f. Make all the steps clear.

(c)   Here, another read-off from the graph is required.

(d)   (i)   There are two points to bring out in your answer, one relating to the meaning of "longitudinal", and one to "progressive".
    (ii)   Part (b) was a "**show that**" question where the answer was given. You must use this value even if you could not arrive at the answer. The problem involves a use of $v = f\lambda$.

Which areas of the syllabus is this question taken from?

• This question requires knowledge of "Kinematics of simple harmonic motion (SHM)" (topic 4.1) and "Wave properties" (topic 4.5).

This answer achieved 3/13

There is no credit here as the answer relates the situation to a spring. There is no need to, and it leads nowhere.

Part of the solution requires the relationship between f and ω. There is credit for this.

The numerical answer is correct, but the student forgets the unit and is penalized accordingly.

The answer is a solution that shows manipulation of the equation, substitution into it, and a final answer.

(a)   1.   The object is spring. $X \in \{0.6; 0\}$ means that after tension spring goes at the position "D"   0

(b) $f = \dfrac{1}{T}; \omega = \dfrac{2\pi}{T} \Rightarrow T = \dfrac{2\pi}{\omega} \Rightarrow f = \dfrac{\omega}{2\pi}$ 1

(c) $A = 10.6$ 0

(d)   (i)   0

    (ii) $v = \lambda f \Rightarrow \lambda = \dfrac{v}{f} = \dfrac{330}{350} = 0.94\ m$ 2

This answer achieved 4/13

The fact that the graph goes through the origin does not necessarily mean that it is a straight line. That is what needs to be seen here.

There is no mention of **energy** propagation. However, the idea of the medium moving in the same direction as that of energy propagation is just about present in the answer.

Try to give answers that are more complete than this. The important points to link are: the direction of energy propagation and the direction in which the medium or the particles in the gas move.

(a) 1. The $|$maximum acceleration$| = |$maximum deceleration$|$ while the $|$positive displacement$| = |$negative displacement$|$

2. The graph passes through the origin at $a = 0$ which is the midpoint between max acc. and max dec. **0**

(b) $f = \dfrac{1}{T}$ **0**

(c) 0.6 mm **1**

(d) (i) A longitudinally progressing wave is one that oscillates in its direction of propagation **1**

(ii) $v = f\lambda \Leftrightarrow 330 = 350 \times \lambda \Leftrightarrow \lambda = \dfrac{330}{350}$
$\lambda = 0.943\ m$ **2**

 This answer achieved 10/13

Even this answer, which gains 2 marks, is not well written. The required arguments are: the line is straight and through the origin, so acceleration is directly proportional to displacement—this is one condition for SHM; a is directly proportional to $-x$, so acceleration and displacement are always in opposite directions—this is the second condition.

This is a well-expressed solution with all the relevant points present.

The meaning of "longitudinal" is here with clear references to the direction in which energy moves and the direction of oscillation of the particles.

The student gives only a weak description of what "progressive" means. A phrase such as "a progressive wave transfers energy by means of vibrations or oscillations" is required.

(a) 1. the displacement from equilibrium position (0,0) is proportional to the acceleration of the object as the graph is a straight line with constant gradient

2. the acceleration is always directed towards equilibrium position. At equilibrium position, acceleration is zero, which is why the graph goes through the origin **2**

(b) $\omega^2 = (-)\dfrac{a}{x}$

$\omega^2 = \dfrac{2900}{0.60 \times 10^{-3}}$

$\omega^2 = 2\pi f$

$f = \sqrt{\dfrac{2900}{0.60 \times 10^{-3}}}$ **4**

(c) 0.6 mm **1**

(d) (i) a wave in which the particles continuously oscillate parallel to the direction of energy transfer, forming periodical rarefractions and compressions. **1**

(ii) $v = f\lambda$
$330 = 350\,\lambda$
$\lambda = 0.94 \cong \underline{1\ m}$ **2**

10. Electric currents, electromagnetic induction and digital technology

Key terms for this chapter

- potential difference—the amount of energy transferred per unit charge that flows through a component
- electric current—the rate at which charge flows in an electric circuit
- resistance—ratio of the potential difference across a material to the electric current in it
- resistivity—a property of a material that causes resistance to current
- emf and internal resistance
- analogue and digital
- capacitance, quantum efficiency and magnification
- charge-coupled device
- pixel
- bit and byte

Electric potential difference, current and resistance

You should know:

- the definitions for:
  - electric potential difference
  - the electronvolt
  - electric current
  - resistance
- what is meant by resistivity
- Ohm's law.

You should be able to:

- solve problems involving:
  - electric potential difference, current and resistance, and power dissipation in resistors
  - changes in potential energy when charges move between points at different potentials
- compare ohmic and non-ohmic behaviour.

Example

Two rectangular blocks, X and Y, of the same material have different dimensions but the same overall resistance. Which of the following equations is correct?

A. resistivity of X × length of X = resistivity of Y × length of Y

B. $\dfrac{\text{length of X}}{\text{cross-sectional area of X}} = \dfrac{\text{length of Y}}{\text{cross-sectional area of Y}}$

C. resistivity of X × cross-sectional area of X = resistivity of Y × cross-sectional area of Y

D. $\dfrac{\text{length of X}}{\text{cross-sectional area of Y}} = \dfrac{\text{length of Y}}{\text{cross-sectional area of X}}$

Answer: B

The usefulness of the resistivity of a material is that its value does not depend on the size or shape of the material. One piece of pure copper will have the same resistivity as another piece of pure copper with a different shape. The phrase "of the same material" tells you that the resistivities are the same. Write down the equation that relates

Electric potential difference, current and resistance (continued)

resistance to resistivity, length and cross-sectional area, and do this for both X and Y. The two resistances are equal and the resistivities will cancel from the equation.

Be prepared

- Develop a clear understanding of what is happening in a simple electric circuit. Charge is flowing through the circuit; this charge is in the form of electrons if the components are metals. The rate at which this charge flows past a point in the circuit is the electric current.

- The concept of potential difference (pd) links strongly to electric field theory. Pd is a measure of the amount of energy transferred when a unit of charge flows through a component. If a lamp gains 15 joules of energy when 2 coulombs of charge flow (whatever length of time this takes), then there is a pd of 7.5 V across the device.

- Linking the energy transfer to charge rather than to time (which we do in mechanics) makes sense. A circuit left on for a very long time may transfer much larger total quantities of energy than another circuit that is only switched on briefly. But if they both have the same pd, we will have a good understanding of the energy transfers.

- The definition of resistance ($R = V/I$ in symbol form) is not the same as Ohm's law. This is a common misunderstanding that you should avoid. Ohm's law says no more than that the potential difference (in an ohmic conductor) is directly proportional to the current in the conductor while the other conditions remain constant.

- You are required to know the different I–V (current–pd) characteristics of an ohmic resistor and a filament lamp. Remember that you could be asked to draw the graph with V plotted on either the x- or the y-axis.

- The *Physics data booklet* contains a comprehensive range of equations relating to electric currents, including the variants of power = current × potential difference. Try to avoid confusing power and energy in this topic. Power is the rate of energy dissipation in joules per second. Energy is the total energy dissipated or transferred in joules.

Electric circuits

You should know:
- the definition for electromotive force (emf)
- what is meant by internal resistance
- what is meant by a potential divider.

You should be able to:
- draw circuit diagrams using accepted circuit symbols
- solve problems involving resistors in series and parallel
- describe how ideal ammeters and voltmeters are used to make measurements in circuits
- explain how sensors such as LDRs, thermistors and strain gauges are used in potential divider circuits
- solve problems involving electric circuits.

Example

Two 6 Ω resistors are connected in series with a 6 V cell. A student incorrectly connects an ammeter and a voltmeter as shown below:

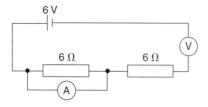

The readings on the ammeter and on the voltmeter are

	Ammeter reading / A	Voltmeter reading / V
A.	0.0	0.0
B.	0.0	6.0
C.	1.0	0.0
D.	1.0	6.0

Answer: B

Although the question does not say so, assume that the cell has zero internal resistance, that the ammeter has zero resistance and that the voltmeter has a very high resistance (infinite, ideally). Placing a 0 Ω ammeter across a resistor effectively shorts the resistance out and it will play no part in the behaviour of the circuit. The voltmeter, on the other hand, has by far the largest resistance of any component in the circuit and will limit the charge flowing around. So the circuit becomes essentially two resistors in series (one 6 Ω, one with a very large resistance) forming a potential divider.

Electric circuits (continued)

Be prepared

- Combining resistors in series is straightforward: simply add the values of the resistors.

- Combining resistors in parallel requires more care. Add the reciprocals of the individual resistance values

$$\frac{1}{R_1} + \frac{1}{R_2} + \dots$$

  Most people can manage this easily enough. What many forget is that the final step in the calculation is to take a final **reciprocal** of the sum to form the final value.

- Electromotive force (emf) is the term used when energy is converted **into** an electrical form from another form. Microphones, photovoltaic cells and dynamos would all be said to have an emf and their voltage output would be termed an emf. Potential difference (pd) is the term used when the conversion is in the opposite direction—**from** electrical to another form. So loudspeakers, filament lamps and electric motors will all have the voltage across them described as a pd.

- Internal resistance arises in many types of component. You will meet it mainly in an electric cell or a battery. The cell requires some of its own supply of energy to allow charge to flow through the internal resistance. The problem is that this pd is "invisible" in that it never appears on a voltmeter. The energy is often described as "lost volts".

- A potential divider circuit is a convenient way to control voltage. You need to have confidence in designing potential divider circuits and in carrying through calculations involving them.

- **Ammeters** are always placed in **series** in the circuit the current of which they are measuring (the charge has to flow through them for it to be measured). **Voltmeters** are always placed in **parallel** with the component for which the potential difference is required.

Analogue and digital signals

You should know:

- different means of information storage in both analogue and digital forms.

You should be able to:

- solve problems where there is conversion between binary and decimal numbers
- explain how information is recovered from CDs
- calculate the depth for a pit on a CD knowing the wavelength of the laser light used
- solve problems of data storage capacity for both CDs and DVDs
- discuss the advantages of storing information in digital form
- discuss the implications for society of ever-increasing data storage capability.

Example

In binary, the decimal number 20 is represented as

A. 11101
B. 11000
C. 10100
D. 00010

Answer: C

The least significant bit (the digit on the right) corresponds to 1, one place to the left of that represents 2, one place to the left of that represents 4, and so on. Add all the numbers together to arrive at the number in denary (in base 10)—the normal number scale.

Be prepared

- Information stored digitally is in the form of "0"s and "1"s. It includes the information contained in CDs, DVDs, MP3 players, iPods and so on. The advantages of this type of storage include the possibility of compression techniques, simplicity, the ability to remove noise from transmitted signals, and so on.

- Analogue information is stored in the form of continuous variables, for example as voltages or magnetic fields. Vinyl records, older tape recorders and photographs recorded on film are all examples of analogue storage. It is difficult to recover the original signal from noise completely in these types of storage.

- Make sure that you can label a diagram of the cross-section of a CD or DVD. The reflection of the laser light at the surface leads to destructive interference or constructive interference depending on whether the reflection takes place at the bump or the flat.

- At the time of writing, 1.5 TB hard disks are common. Predictions indicate that, by 2020, 15 TB hard disk units in domestic computers will be commonplace, assuming that another technology has not taken over in the meantime. You should be prepared to write about the advantages and disadvantages for society of this increasing reliance on digital data.

Data capture: digital imaging using charge-coupled devices (CCDs)

You should know:

- the definition of:
  - capacitance
  - quantum efficiency of a pixel
  - magnification
- the structure of a CCD
- how light incident on a pixel causes charge to build up
- how the image on a CCD is digitized
- the criterion for resolution on a CCD.

You should be able to:

- discuss how the quality of the processed image depends on:
  - quantum efficiency
  - magnification
  - resolution
- describe a range of practical uses of a CCD
- list the advantages of CCDs when compared with the use of film
- outline how the image stored in a CCD is retrieved
- solve problems involving the use of CCDs.

Example

A CCD camera is used to take a picture of an object of length 30 m. The image of an object on the chip measures 0.03 mm. Which of the following gives the magnification of the camera?

A. 10^6

B. 10^3

C. 10^{-3}

D. 10^{-6}

Answer: D

The definition of magnification for a CCD is the ratio of the length of the image on the CCD to the length of the object. Use this relationship to calculate the magnification. In this and all questions like it, take care to calculate the magnification, not its reciprocal, and be careful with the units of length in the question!

Be prepared

- There are many uses for a CCD beyond the straightforward application of a digital still camera. Ensure that you have noted down and learned some of these uses.

- As the size of the smallest pixels that can be manufactured shrinks, so CCDs are able to rival analogue (film) techniques more and more. The digital technique also has advantages in terms of storage, portability and manipulation.

- Capacitance is the ratio of the charge stored on the pixel to the pd at which it is stored. As more electrons are ejected from the pixel as the photons arrive, the charge builds up. The pd is directly proportional to this charge through the capacitance relationship. So, the pd is directly proportional to the number of photons arriving, in other words, to the brightness of the object.

- Not all photons will eject an electron. The quantum efficiency is a measure of the number of photons that achieve this compared to the total number arriving.

- Be prepared to draw and label a diagram of a CCD or an individual pixel.

- After an exposure is completed, every pixel has a potential difference across it. These voltages are read off the pixels and converted into digital values that are stored in the memory of the device together with the location of the pixel on the sensor.

- The criterion for resolution on a CCD is that two points on an object are just resolved if the central maxima of the diffraction patterns of the points fall two pixels apart, with a dip in intensity at the pixel between them.

Part 2   This question is about electric circuits.

The components shown below are to be connected in a circuit to investigate how the current I in a tungsten filament lamp varies with the potential difference V across it.

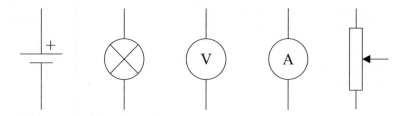

(a)   Construct a circuit diagram to show how these components should be connected together in order to obtain as large a range as possible for values of potential difference across the lamp.   *[4]*

(b)   On the axes, sketch a graph of I against V for a filament lamp in the range $V=0$ to its normal working voltage.   *[2]*

(c)   The lamp is marked with the symbols "1.25 V, 300 mW". Calculate the current in the filament when it is working normally.   *[1]*

(d)   The resistivity of tungsten at the lamp's working temperature is 4×10^{-7} Ω m. The total length of the tungsten filament is 0.80 m. Estimate the radius of the filament.   *[4]*

(e)   The cell is connected to two identical lamps connected in parallel. The lamps are rated at 1.25 V, 300 mW. The cell has an emf of 1.5 V and an internal resistance of 1.2 Ω. Determine whether the lamps will light normally.   *[4]*

[Taken from SL paper 2, November 2009]

How do I approach the question?

(a) There are two possibilities: a circuit with a variable resistor; and a circuit with a potential divider. One of these will give a wider range of pd values. When drawing the circuit, take care that you use the standard symbols for components, and also that the ammeter and voltmeter are connected in the correct positions.

(b) You are required to know the shape of the *I–V* curve (characteristic) for an ohmic conductor and a lamp. The ohmic conductor is straightforward, the lamp less so. But if you remember that, as a lamp heats up (at higher currents), the resistance increases, you should be able to work out the shape of the graph if you have forgotten it.

(c) Use the equation that connects power, pd and current to calculate the current. Take care with the power of 10, as the power is expressed in the question as milliwatts.

(d) This is an estimate, not—in this case—because you are required to make an educated guess at one or more of the data values, but because we cannot be entirely certain about the value of the answer. Can you think why this is?

(e) This is a "**determine**" question at objective group 3 level, in which you are required to make your answer clear and well structured. There are a number of ways to tackle the question. The question requires a judgment of the extent to which the internal resistance of the cell will affect the current in the circuit. Begin by calculating the total resistance of the lamps in parallel and then include the internal resistance into this total. This will enable you to calculate the current in the lamps and then to assess whether this will enable them to be at normal brightness.

Which areas of the syllabus is this question taken from?

- This question requires knowledge of "Electric potential difference, current and resistance" (topic 5.1) and "Electric circuits" (topic 5.2).

This answer achieved 6/15

 The ammeter and voltmeter are placed in the correct positions relative to the measurements they are intended to take.

 The variable resistor is not connected correctly. It will give a variation in pd values, but will not give the largest range possible. For this, the resistor must be connected as a potential divider.

The student has the right idea but has confused the axes. This graph indicates that the resistance of the lamp becomes smaller as the current (temperature) of the lamp increases.

The graph must go through the origin and be straight at that point.

This is a straightforward calculation competently laid out with all detail present.

The solution begins with the correct equation and a correct change to include the radius of the filament. However, in the third line, the student omits the resistivity. Take care when rearranging equations that you do not lose terms in this way.

The assumption that the current is 0.24 A is incorrect. The cell will require energy if there is to be charge flowing through it. This will reduce the current below the value calculated in part (b).

(a)

3

(b)

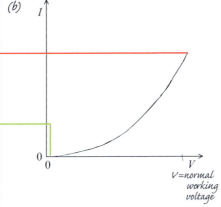

V = normal working voltage

1

(c) $P = VI$

$I = \dfrac{P}{V} = \dfrac{300 \times 10^{-3}\,W}{1.25V} = 0.24A$

1

(d) $R = \dfrac{\rho L}{A}$

$R = \dfrac{\rho L}{\pi r^2}$

$R = \dfrac{L}{\pi r^2}$ if density is the same/constant

$r^2 = \dfrac{L}{\pi \times R}$

$\therefore r = \sqrt{\dfrac{0.80}{\pi \times 4.0 \times 10^{-7}}} = 1.25 \times 10^{-3}\,m$

0

(e) Using $\varepsilon = I(R + r)$ to find R of lamps

$1.5 = 0.24\,(R + 1.2)$

$1.5 = 0.24R + 0.288$

$R = 5.05$

The resistance of the lamps is much higher than the emf so the lamps will not light normally and the cell has a larger internal resistance.

1

This answer achieved 9/15

This student, like the previous one, fails to draw a circuit that can produce the largest possible range of pd values

A straight line is appropriate close to the origin, but a curve will be expected at the pd that corresponds to the normal working voltage. It is helpful in sketching a graph to mark any relevant points such as the normal working voltage.

This is a correct solution marred by an incorrect unit.

This is a good example of how to set out a calculation neatly and clearly.

Again, the assumption of the current value is incorrect.

(a)

3

(b)

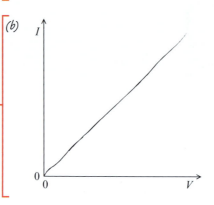

1

(c) $P = IV$

$300 \times 10^{-3} = I \times 1.25$

$I = 0.24C$

0

(d) $R = \dfrac{V}{I} = \dfrac{1.25}{0.24} = 5.2$

$R = \dfrac{\rho L}{A} = 5.2 = \dfrac{4 \times 10^{-7} \times 0.8}{A}$

$A = 6.14 \times 10^{-8}\ m^2 = \pi r^2$

$\therefore r = 1.4 \times 10^{-4}\ m$

4

(e)   Yes they will light.

$\varepsilon = IR + Ir = 1.5 = 0.24R + 0.24 \times 1.2$

$1.212 = 0.24R$

$R = 5.05$

$\dfrac{1}{R_T} = \dfrac{1}{5.2} + \dfrac{1}{5.2} + \dfrac{1}{5.2} = \dfrac{15}{26}$ $R = \dfrac{26}{15} = 1.73$

1

This answer achieved 15/15

This diagram shows that the student has recognized that a potential divider is needed in this question.

Make sure that you understand how the potential divider operates, and how it should be connected in the circuit.

(a)

4

(b)

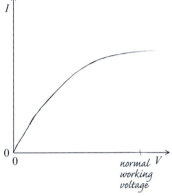

2

(c) $I = \dfrac{P}{v}$

$= \dfrac{300 \times 10^{-3}}{1.25}$

$= 0.24A$

1

(d) $R = \dfrac{V^2}{P} = 5.2\,\Omega$

$R = \dfrac{\rho L}{\pi r^2}$

$r^2 = \dfrac{4 \times 10^{-7} \times 0.80}{\pi \times 5.2}$

$r = 1.398 \times 10^{-4}$

$r = 1.4 \times 10^{-4}\,m$

4

This is a good solution with a calculation of the resistance of the two lamps in parallel and includes the internal resistance in series. This allows a calculation of the new current. As this is less than the original current, the student correctly deduces that the lamps cannot light normally.

(e) $\dfrac{1}{R_T} = \dfrac{1}{5.2} + \dfrac{1}{5.2}$

$R_T = 2.6\,\Omega$

$I = \dfrac{1.5}{2.6 + 1.2} = 0.394A$

$Current\ for\ each\ lamp = \dfrac{0.394}{2} = 0.197A < 0.24A$

The lamps will not light normally because the current is too low.

4

A5. This question is about charge-coupled devices (CCD).

(a) With reference to a CCD, state what is meant by a pixel. *[2]*

(b) Outline how light falling on a CCD leads to an electrical signal being produced by a pixel. *[3]*

(c) State **one** other piece of information that needs to be collected, in addition to the electrical signal in (b), in order that an image may be formed. *[1]*

(d) Suggest **two** advantages of a CCD in comparison with a photographic film for image production. *[2]*

1.

2.

[Taken from HL paper 2, time zone 2, May 2009]

How do I approach the question?

(a) There are 2 marks and therefore two points to make. The question begins "with reference to a CCD", so you must make this connection clear.

(b) There are 3 marks. So you need to give a clear description that includes the arrival of a photon at the pixel, the subsequent ejection of the electron, and the consequence for the pixel.

(c) The camera needs more information than just the pd from each pixel. It also needs to know where the pixel is in the sensor.

(d) In a question such as this, remember that this is a physics exam and whatever you suggest should be, as far as possible, referring to the physics of the situation.

Which areas of the syllabus is this question taken from?

- This question requires knowledge of "Data capture; digital imaging using charge-coupled devices (CCDs)" (topic 14.2).

This answer achieved 2/8

There is no clear description of a pixel, and the answer veers into saying what a CCD is. There is no clear distinction between pixel and CCD.

There is some credit given for the idea that the photon brings energy with it.

The false idea that the pixel "lights up" is certainly enough to disqualify that part of the answer.

The first suggestion suggests the re-usable nature of the sensor in a digital camera.

The second suggestion was simply not true at the time the answer was written.

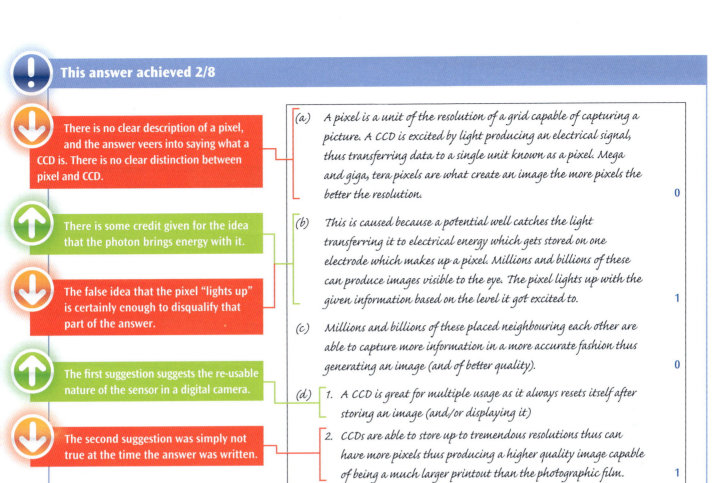

(a) A pixel is a unit of the resolution of a grid capable of capturing a picture. A CCD is excited by light producing an electrical signal, thus transferring data to a single unit known as a pixel. Mega and giga, tera pixels are what create an image the more pixels the better the resolution.   0

(b) This is caused because a potential well catches the light transferring it to electrical energy which gets stored on one electrode which makes up a pixel. Millions and billions of these can produce images visible to the eye. The pixel lights up with the given information based on the level it got excited to.   1

(c) Millions and billions of these placed neighbouring each other are able to capture more information in a more accurate fashion thus generating an image (and of better quality).   0

(d) 1. A CCD is great for multiple usage as it always resets itself after storing an image (and/or displaying it)

2. CCDs are able to store up to tremendous resolutions thus can have more pixels thus producing a higher quality image capable of being a much larger printout than the photographic film.   1

This answer achieved 4/8

"One unit of area" does not mean anything—perhaps "smallest unit of area" might have been better. The resolution point is irrelevant in this answer as it does not refer to the CCD.

This answer begins well. There is, however, no reference to the increase in potential as charge is lost (through a capacitive effect). So the third mark is lost. There is no current in this case, as the resultant charge on the pixel does not move. So this sentence is incorrect.

The intensity of the electrical signal is the information that we already know is collected.

Two reasonable advantages are well expressed.

(a) A pixel is one unit of area on the CCD and the amount (size) of pixels determines the maximum resolution possible.   0

(b) As light falls on a CCD, electrons are released (due to the photoelectric effect) from the metal on the pixel of the CCD. As more light falls, more electrons are released. This causes a current, and varying amounts of photoelectrons emitted result in varying electrical signals.   2

(c) The intensity of the electrical signal (i.e. the intensity of the incident light).   0

(d) 1. Images produced from a CCD are less easily corrupted as those on photographic film

2. CCD's provide easier manipulation of images (they can be viewed and deleted before printing).   2

This answer achieved 6/8

Again, the answer does not make clear what is meant by a unit, but the reference to the CCD being a semiconductor scores a mark.

This answer focuses on the photoelectric effect and the charge separation that occurs in the pixel. Like the previous answer, it fails to explain that the build-up of charge leads to an increase in potential difference across the pixel.

The mark is just given here. The best answer would be "the location of the pixels" but this just conveys the meaning of where the pixels are relative to each other.

(a)   A pixel is one unit of an image formed by a CCD which is made from a semiconductor   1

(b)   As light falls on a CCD each unit in the CCD emits photoelectrons so that the amount depends on the intensity of light falling on the CCD. The electrons emitted create a charge that is left on the pixel.   2

(c)   Order of the pixels   1

(d)   1.   Easier to edit the image

      2.   Can be used in less light as photo efficiency of CCDs is higher than with photographic film   2

The first suggestion is just acceptable as being related to physics. The second advantage is a good, solid piece of physics, and this may have swayed the examiner to give a benefit of the doubt with the first part.

11. Fields and forces

Key terms for this chapter

- gravitational field, gravitational field strength, gravitational potential and gravitational potential energy
- electric field, electric field strength, electric potential and electric potential energy
- magnetic field, magnetic flux and magnetic flux linkage
- rms and peak values for sinusoidal currents
- transformer

Gravitational force and field, gravitational potential and gravitational potential energy

You should know:

- Newton's universal law of gravitation
- the definition of:
  - gravitational field strength
  - gravitational potential
  - gravitational potential energy
- a mass has a gravitational potential of zero when it is at infinity relative to another object
- that "infinity" is a place so far away that the ratio

$$\frac{1}{\text{distance to the object}}$$

  is effectively zero

- escape speed is the speed required at the surface of a planet to escape completely from the gravity field of the planet (that is, to reach infinity)
- gravitational field strength is numerically equal to the gravitational potential gradient in a gravitational field
- that no work is needed to move around on an equipotential surface
- equipotential surfaces are always at 90° to the field lines.

You should be able to:

- derive and use the equation for gravitational field strength at the surface of a planet
- solve problems using gravitational forces and fields
- determine the gravitational potential of a combination of point masses
- sketch equipotential patterns around point masses
- solve problems involving:
  - gravitational potential and potential energy
  - escape speed, including being able to derive the equation for escape speed.

Gravitational force and field, gravitational potential and gravitational potential energy (continued)

Example

The magnitude of the gravitational field strength at the surface of a planet of radius R is $8.0\,\text{N}\,\text{kg}^{-1}$. Which of the following is a correct expression for the gravitational potential at the surface of the planet?

A. $-\dfrac{8.0}{R}$

B. $-\dfrac{8.0}{R^2}$

C. $-\dfrac{R}{8.0}$

D. $-8.0R$

Answer: D

There are at least two separate ways to attempt this question (as is so often the case with a multiple-choice question). The first way is through the physics: knowing the equation for the gravitational field strength of a planet

$$\frac{GM}{R^2}$$

and the gravitational potential equation will allow you to decide what the gravitational field strength needs to be multiplied by in order to become the potential. The second way is to think about the units: you are told that the field strength is in $N\,kg^{-1}$; you should remember that potential is measured in $J\,kg^{-1}$.

Be prepared

- It is important to understand the links between the quantities such as potential, potential energy, field strength and force. What is written in this "Be prepared" section applies equally to electric fields and forces, and you should draw out the parallels between the two areas of the syllabus.

- One way to revise this topic is to look at the relationships between gravitational force and gravitational field strength, and between gravitational potential energy and gravitational potential.
  - Gravitational potential = gravitational potential energy per unit mass
  - Gravitational field strength = gravitational force per unit mass

- Another way is to concentrate on a more mathematical approach.
  - gravitational field strength
    $$= \frac{\text{change in gravitational potential}}{\text{change in distance}}$$
  - gravitational force
    $$= \frac{\text{change in gravitational potential energy}}{\text{change in distance}}$$

Electric field, electric potential and electric potential energy

You should know:

- that there are two types of electric charge

- charge is conserved

- Coulomb's law

- the definition of:
  - electric field strength
  - electric potential
  - electrical potential energy

- electric field strength is numerically equal to the electric potential gradient in an electric field

- no work is needed to move around on an equipotential surface

- equipotential surfaces are always at 90° to the electric field lines.

You should be able to:

- determine the electric potential of a combination of point charges

- sketch equipotential patterns around point charges

- solve problems involving electric potential and potential energy

- describe and explain the difference in the electrical properties of conductors and insulators

- draw electric field patterns for charge configurations.

Example

The two graphs below represent the variation with distance, d, for $d = r$ to $d = 2r$, of the electric field and the electric potential around an isolated point charge.

Electric field, electric potential and electric potential energy (continued)

Graph 1:

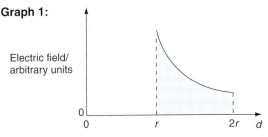

Electric field/ arbitrary units

Graph 2:

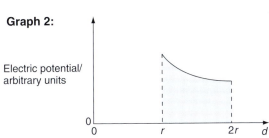

Electric potential/ arbitrary units

The work done by an external force in moving a test charge $+q$ from $d = 2r$ to $d = r$ is equal to q multiplied by the

A. shaded area under graph 1.
B. shaded area under graph 2.
C. average value of the electric field.
D. average value of the electric potential.

Answer: A

The key here is to remember the rule that work done = force × distance. Although this rule only applies to a constant force, it helps us to remember that the area under a graph of force plotted against distance will give the value for the work done. Here, however, neither graph is force against distance. You need to choose which area will give you the work done per unit charge. The work done will then be the magnitude of the charge itself multiplied by your answer.

Be prepared

- Many of the ideas in the "Be prepared" section for gravitational fields apply here. In particular, once you have mastered the connected concepts of field, force, potential and potential energy in one area of the syllabus, you have mastered it in the other area too.

- Another concept shared between gravitation and electrostatics is that of the field line and the equipotential.
  - Field lines are the directions along which a positive charge will move (in other words, away from another positive charge and towards a negative charge).
  - Equipotentials are lines along which a charge moves without doing any work or having any work done on it.

- The fact that there are two types of charge labelled +ve and –ve leads to more complication in electrostatics than in gravitation. Take potential, for example. In general, potential in gravitation has a negative sign if infinity is the zero point. Potential takes either sign in electrostatics depending on the signs of the charges concerned.

Magnetic force and field

You should know:

- that magnetic fields arise from the movement of charge
- the definition of the magnitude and the direction of a magnetic field.

You should be able to:

- draw the magnetic field patterns due to electric currents
- determine the direction of the force that acts on a current-carrying conductor in a magnetic field
- determine the direction of the force that acts on a charge moving in a magnetic field
- solve problems that involve magnetic forces, fields and currents.

Example

A current-carrying wire is in the same plane as a uniform magnetic field. The angle between the wire and the magnetic field is θ.

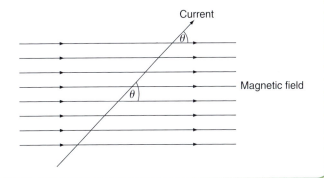

Magnetic force and field (continued)

The magnetic force on the current-carrying wire is

A. zero.

B. into the plane of the paper.

C. out of the plane of the paper.

D. at an angle θ to the direction of the magnetic field.

Answer: B

This question involves the use of a direction rule that connects the current in a wire, the magnetic field in which the wire sits, and the force that acts on the wire as a result. There are several possible rules that you may have learnt, including Fleming's left-hand rule. There is the added complication that the current is not perpendicular to the field direction. The current can be imagined as having two components: parallel to the field and perpendicular to it. One of these will have no influence on the force. Decide which component is important, and use the rule you know to work out the direction of the force.

Be prepared

- You can be asked to determine the direction of the magnetic force acting on both a moving charge and a current-carrying wire. These are the same skill because a charge moving in a wire and a current in a wire are one and the same thing. Always, with your direction rules, use the direction in which a **positive** charge would move.

- Units: the unit of magnetic field is the tesla (T). This can be derived from the formula $F = BIl\sin\theta$. Rearranging the units indicates that T can also be written $N\,A^{-1}\,m^{-1}$.

Induced electromotive force (emf)

You should know:

- the definitions of magnetic flux and magnetic flux linkage

- Faraday's law

- Lenz's law.

You should be able to:

- describe how an emf is induced when there is relative motion between a conductor and a magnetic field

- describe how an emf is induced when a magnetic flux changes with time

- derive the formula for the emf induced in a straight conductor moving in a magnetic field

- solve electromagnetic induction problems.

Example

A metal ring has its plane perpendicular to a magnetic field.

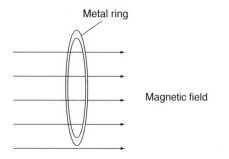

Metal ring

Magnetic field

The magnetic flux through the ring increases at a constant rate by 4.0×10^{-5} Wb in 5.0 s. During this change, the emf induced in the ring

A. remains constant at $8\,\mu V$.

B. remains constant at $20\,\mu V$.

C. increases from zero to $8\,\mu V$.

D. increases from zero to $20\,\mu V$.

Answer: A

Faraday's law states that the emf induced in a conductor is equal to the rate of change of flux linkage. To answer a question such as this, you need first to know this rate of change. Complications could include the number of turns in the coil, but in this case the ring has only one turn. You need to take the total change of flux linkage and divide it by the time taken for the change. The final decision is whether the emf is constant or changing. In this case you are told that the rate of change of flux linkage is constant. So Faraday's law indicates that the emf is also constant.

Be prepared

- Whenever a conductor moves relative to a magnetic field, or whenever the field that links a conductor varies, an emf will be induced in the conductor. Even though an emf is generated, this does not necessarily give rise to a current—that depends on whether there is a complete conducting circuit, for example. The emf generation is the principal effect.

Induced electromotive force (emf) (continued)

- If the movement or the field change occurs in a shorter time, the emf will be larger.
- The direction in which the emf is produced will oppose the change that is causing it. This is Lenz's law.
- Flux linkage is measured in weber turns (amount of flux × number of turns in the coil) and so rate of change of flux linkage is weber turns per second ($Wb\,s^{-1}$) and this is dimensionally the same as a volt.
- Flux density in webers per square metre ($Wb\,m^{-2}$) is numerically the same as the magnetic field strength, which is normally measured in tesla (T). The magnetic field strength may be thought of as the number of field lines per square metre, and so $1\,T = 1\,Wb\,m^{-2}$. You can use either unit for magnetic field strength—they are both correct.
- The question above required you to use Faraday's law. Mathematically, this can be expressed as

$$\mathcal{E} = -\frac{N\,d\Phi}{dt}$$

where N is the number of turns in the coil in which the emf is induced, Φ is the flux linkage and t is the time. The negative sign reminds us again of Lenz's law. If the system has the opportunity to do so, it will try to reduce the effect of the change that is producing the emf.

Alternating current (ac)

You should know:

- how an ac generator operates
- what is meant by the root mean squared (rms) value of an alternating current or voltage
- the relationship between peak and rms values for sinusoidal voltages and currents
- how an ideal transformer operates.

You should be able to:

- describe the emf induced in a coil rotating in a uniform magnetic field
- describe how the emf induced in an ac generator changes when the generator frequency changes
- solve problems using peak and rms values
- solve problems for ohmic resistors in ac circuits
- solve problems involving ideal transformers.

Example

The maximum output voltage of a generator is V_0. The frequency of rotation of the generator coil is doubled. Which of the following gives the new maximum output voltage?

A. V_0

B. $\sqrt{2}\,V_0$

C. $2V_0$

D. $4V_0$

Answer: C

Doubling the rate of rotation of the coil in an electrical generator has two effects. The most obvious change is that the output

frequency doubles. But Faraday's law has a part to play too! The flux linked goes from zero to a maximum value in half the time. What change does this make to the rate of change of flux linkage, and how does this change the induced emf?

Be prepared

- Alternating current (ac) is used for power transmission because it can be generated, transformed and transmitted relatively easily.
- The ac waveform is sinusoidal. Important values associated with it are the frequency f, the peak value of the current or voltage, and the root mean squared (rms) value of the current I or voltage V. The rms values are needed because the mean (average) values of I and V over one cycle are both zero (the values are both positive and negative over one cycle). The way around the problem is to square all the values for one cycle (this makes them all positive), then take the average or mean, and then take the square root in order to get the values back into the correct units. Hence: the (square) root (of a) mean (of) squared (values).
- Electrical energy can be generated in ac machines in which a coil is rotated inside a magnetic field. Make sure that you understand:
  - the component parts of the generator
  - the way in which current is removed from the coil using slip rings
  - how the shape of the emf–time graph relates to the position of the coil
  - why the graph has this shape (why the emf takes its maximum value at a particular point in the cycle).

Alternating current (ac) (continued)

- If the rate at which the coil rotates is changed, what change does this make to the output of the ac generator? There are two changes: one to the value of the peak (maximum) voltage, and one to how often this maximum occurs. Make sure that you mention both changes and the size of the change.

- A transformer is used to change the values of I and V. The syllabus assumes that the transformers are ideal and have an efficiency of 100% with no energy losses.

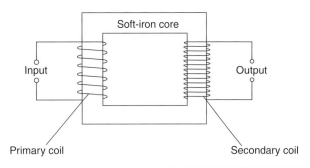

Soft-iron core

Input

Output

Primary coil

Secondary coil

- Revise ideas about the transformer by breaking the operation down into a series of steps.
  - Alternating current I_p at pd V_p is supplied to the primary coil.
  - This produces an alternating magnetic field inside the coil and therefore inside the soft-iron core.
  - The change of magnetic flux inside the secondary coil leads to the generation of an alternating emf V_s in the secondary coil.
  - If there is a complete circuit in the output, this will produce an alternating current I_s.

- For a transformer:
$$\frac{V_s}{V_p} = \frac{n_s}{n_p}$$
where n_s and n_p are the number of secondary and primary turns.

- For a 100% efficient transformer:
$$V_p I_p = V_s I_s$$

Transmission of electrical power

You should know:

- why there are power losses in transmission lines and real transformers.

You should be able to:

- explain the use of step-up and step-down transformers in electrical power transmission
- solve problems on the operation of real transformers and power transmission
- suggest how currents can be induced in the human body by electromagnetic fields
- discuss the possible risks of living and working near high-voltage power lines.

Example

High voltages are used for the transmission of electric power over long distances because

A. high voltages can be stepped down to any required value.

B. larger currents can be used.

C. power losses during transmission are minimized.

D. transformers have a high efficiency.

Answer: C

Electrical energy can be sent long distances but high current values lead to large power losses (because the power lost in a conductor = $I^2 R$, where I is the current and R is the resistance of the conductor). With the large currents and large transmitted energies that national transmission systems require, the power lost in the cables will be much greater than the power received at the far end from the generator! This is where transformers are helpful, a step-up transformer can make the maximum value of the alternating pd 1000 times larger, and at the same time make the alternating current maximum 1000 times smaller. That means a power loss that is 10^6 times smaller, a useful change indeed with much less loss in the transmission cables. This question is a good example of choosing the most appropriate response. Three of the answers are most definitely true, but one response completes the sentence much better than the others.

Be prepared

- The explanation provided for the question above gives the physics associated with power loss in transmission systems, and you should be ready both to describe this and to answer questions based on power-loss calculations. Power losses in transformers occur as

Transmission of electrical power (continued)

resistive losses in the coils and heating losses in the core due to currents induced in it. The soft-iron core is laminated to help prevent this.

- Step-up and step-down transformers are required in the transmission system. Remember that the term always refers to the **voltage** change, so step-up increases the alternating pd and reduces the current.

- The transformer–transmission line system is made from materials that are good conductors, but, even so, resistance makes its presence felt and power is lost here through heating. Energy is also lost in the transformers themselves because of eddy currents (induced currents

that circulate in the core and are reduced by the use of laminations) and because of other effects such as mechanical vibration caused by the interaction of the magnetic fields and the currents.

- A controversial area of risk at present concerns the presence of high-voltage power lines where people live and work. Such people are therefore subject to low-level alternating electromagnetic fields. Current evidence is that these fields do not harm genetic material in adults. However, there is the suggestion that such damage is possible in children and the evidence is not clear-cut either way.

Part 2   This question is about gravitational fields and potential.

(a)   Define *gravitational field strength* and state how it is related to gravitational potential.   *[3]*

Definition:

Relationship:

(b)   A binary star system consists of two stars with masses M_1 and M_2 rotating about a common centre. The centres of the two stars are separated by a distance $R = 1.2 \times 10^{10}$ m.

The diagram is not to scale.

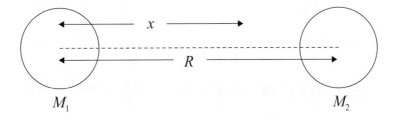

The total gravitational potential due to the stars at any point along a line joining their centres is V. The graph shows how V varies with the distance x from the centre of star M_1. (Values of the potential inside each star are not known.)

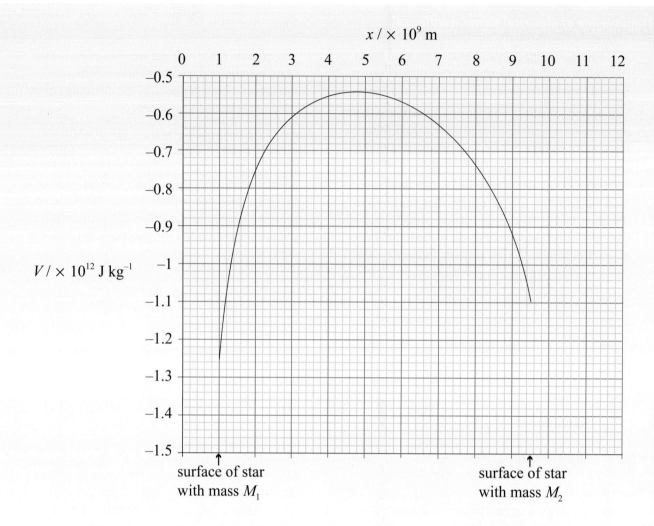

$x / \times 10^9$ m

$V / \times 10^{12}$ J kg^{-1}

surface of star with mass M_1

surface of star with mass M_2

A particle is launched with kinetic energy E_K from the surface of star with mass M_2. The particle arrives at the surface of the star of mass M_1. Use the graph to

(i)   explain whether the kinetic energy of the particle at the surface of M_1 is less than, equal to, **or** larger than E_K.   [2]

(ii) determine the distance x at which the gravitational field strength due to the two stars is zero.   [2]

(iii) determine the ratio $\dfrac{M_1}{M_2}$.   [3]

[Taken from HL paper 2, time zone 1, May 2009]

How do I approach the question?

(a) Define gravitational field strength in terms of the force acting on unit mass. The relationship is shown in the *Physics data booklet*. Identify the appropriate equation and remember to state what the symbols mean.

(b) Look carefully at the graph and analyse what it is telling you.
  (i) There is a difference between the gravitational potentials at the surface of the planets. The particle will require energy to reach the peak of the graph. How much potential does it lose on the way to the second planet and in what form does this appear?
  (ii) The slope of the graph is zero at the peak. What is the relationship between gravitational field strength and the way in which gravitational potential varies with distance?
  (iii) The peak of the graph gives the point at which the two gravitational fields are equal in strength but opposite in direction. So, the magnitude of GM/R^2 is equal for both planets. Deduce the distances of the planet centres from the peak and use the appropriate ratio to determine M_1/M_2.

Which areas of the syllabus is this question taken from?

- This question requires knowledge of "Electric force and field" (topic 6.2) and "Electric field, potential and energy" (topic 9.3).

! This answer achieved 1/10

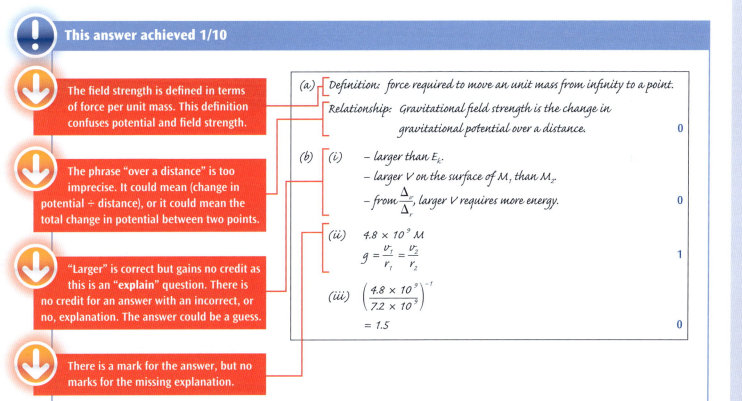

The field strength is defined in terms of force per unit mass. This definition confuses potential and field strength.

The phrase "over a distance" is too imprecise. It could mean (change in potential ÷ distance), or it could mean the total change in potential between two points.

"Larger" is correct but gains no credit as this is an "explain" question. There is no credit for an answer with an incorrect, or no, explanation. The answer could be a guess.

There is a mark for the answer, but no marks for the missing explanation.

(a) Definition: force required to move an unit mass from infinity to a point.
Relationship: Gravitational field strength is the change in gravitational potential over a distance. **0**

(b) (i)
  – larger than E_k.
  – larger V on the surface of M_1 than M_2.
  – from $\frac{\Delta_v}{\Delta_r}$, larger V requires more energy. **0**

(ii) 4.8×10^9 M
  $g = \frac{v_1}{r_1} = \frac{v_2}{r_2}$ **1**

(iii) $\left(\frac{4.8 \times 10^9}{7.2 \times 10^9}\right)^{-1}$
  $= 1.5$ **0**

This answer achieved 3/10

✓ This is correct, but try not to use "gravity". "Gravitational field" is better.

This answer does not deal with the relationship between field strength and potential. It appears to be defining gravitational potential (incorrectly).

This is a poor way to tackle the question. Think in terms of the conversion of potential energy to kinetic energy rather than in terms of speed.

This is a confused answer where the approach is to calculate masses directly. This is not necessary.

(a) Definition: the force exerted by gravity per unit mass.

Relationship: The gravitational potential refers to the theoretical work done when bringing something from infinity to the point of the gravitational field. **1**

(b) (i) it will arrive with less energy at M_1 because it first is decelerated by M_2's pull which is greater than the forward thrust it gets from M_1's pull. Thus over the travel it will be slower at M_1 than its launch speed making $KE_{final} < E_K$ **0**

(ii) At approximately 4.9×10^9 m because the tangent to the curve on the graph at this point is 0. **2**

(iii) $V = \dfrac{-GM_1}{r} \Rightarrow -.91 \times 10^{12} = \dfrac{-GM_1}{1.5 \times 10^9 m} \Rightarrow M_1 = 2.03 \times 10^{31}$ kg

$r_e = 1.2 \times 10^{10} - 1.5 \times 10^9 m = 1.05 \times 10^{10} \Rightarrow$

$-.9 \times 10^{12} = \dfrac{-GM_2}{r^2} \Rightarrow M_2 = 1.42 \times 10^{32}$ kg

$\dfrac{M_1}{M_2} = \dfrac{2.03 \times 10^{31} kg}{1.42 \times 10^{32} kg} = 0.1429 \approx 0.143$ **0**

This answer achieved 7/10

The field strength is related to $\Delta V/\Delta r$, not simply the potential.

This does not gain both marks, as the explanation is not entirely correct. But there was enough understanding to award the mark for the size of E_K.

This is a correct answer and a comment that shows an understanding of the problem.

This answer gives a complete explanation and the physics of the situation is obvious.

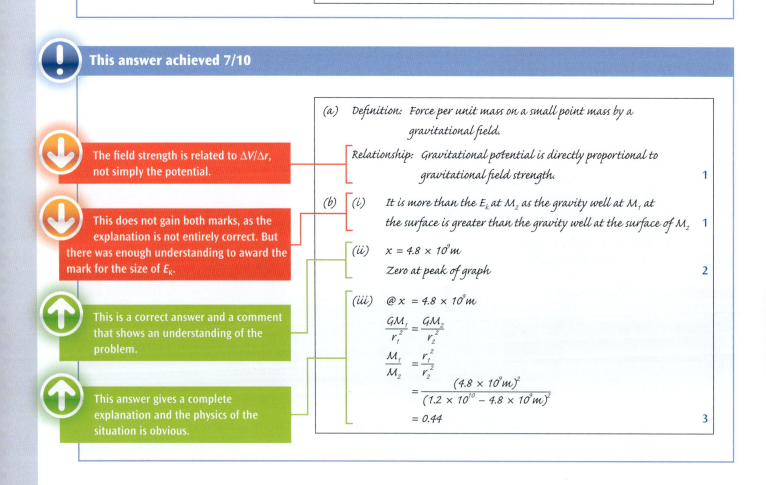

(a) Definition: Force per unit mass on a small point mass by a gravitational field.

Relationship: Gravitational potential is directly proportional to gravitational field strength. **1**

(b) (i) It is more than the E_k at M_2 as the gravity well at M_1 at the surface is greater than the gravity well at the surface of M_2 **1**

(ii) $x = 4.8 \times 10^9$ m

Zero at peak of graph **2**

(iii) @ $x = 4.8 \times 10^9$ m

$\dfrac{GM_1}{r_1^2} = \dfrac{GM_2}{r_2^2}$

$\dfrac{M_1}{M_2} = \dfrac{r_1^2}{r_2^2}$

$= \dfrac{(4.8 \times 10^9 m)^2}{(1.2 \times 10^{10} - 4.8 \times 10^9 m)^2}$

$= 0.44$ **3**

A4.   This question is about alternating current.

(a)   The graph shows the variation with time t of the output voltage V of an ac generator of negligible internal resistance.

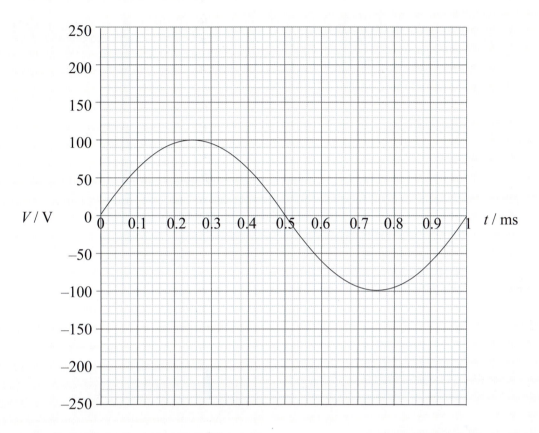

V / V

t / ms

A resistor of resistance 25 Ω is connected across the output of the generator.

Calculate

(i)   the rms value of the current in the resistor. [2]

(ii)   the average power dissipated in the resistor. [1]

(iii)   the power dissipated in the resistor at 0.40 ms. [2]

(b)   The frequency of rotation of the generator coil is now doubled. Sketch, using the axes in (a), the variation with t of the new output voltage V. [2]

[Taken from HL paper 2, time zone 1, May 2009]

How do I approach the question?

(a) (i) The relationship between the peak values of current or pd and the root mean squared (rms) values is that $I_{rms} = I_{peak}/\sqrt{2}$ (or $V_{rms} = V_{peak}/\sqrt{2}$). You will need to use this together with the definition of resistance, which applies at all times for resistors in ac or dc circuits, to calculate the current. Make sure that you read the appropriate values from the graph accurately.

(ii) The average power is equal to the product $I_{rms}V_{rms}$ (or $I_{peak}V_{peak}/2$).

(iii) At an instant in time, the power equals the product of the instantaneous values of current and pd. A careful read-off from the graph is needed. However, you must either calculate the current (in the same way as before) or recall that

instantaneous power $= \dfrac{V_{inst}^{2}}{R}$.

(b) The frequency of rotation is doubled and you need to consider in your answer both of the changes that occur to the waveform. At this level of exam you are expected to be able to sketch a reasonably accurate freehand sine or cosine curve. A few minutes practice in advance of the exams will pay dividends.

Which areas of the syllabus is this question taken from?

• This question requires knowledge of "Induced electromotive force (emf)" (topic 12.1) and "Alternating current" (topic 12.2).

This answer achieved 0/7

 Writing a note to the examiner will not help you gain marks. It would have been better to spend time improving the shape of the sketched graph.

The student has realized that the frequency of the pd has to double but forgotten that the rate of change of flux linkage will have doubled too.

The answer begins well with the correct equation for the definition of resistance. However, the rearrangement of the equation is incorrect and there is no quoted read-off for pd from the graph. There is no unit quoted for the current.

 Always write down what you read from a graph, as it may gain you a mark if there are later errors.

(a) *round like a sine function*

(i) $R = \dfrac{V_{rms}}{I_{rms}}$

$I_{rms} = \left(\dfrac{V_o}{\sqrt{2}}\right).25 = 1.78 \times 10^{3}$ 0

(ii) $P_{av} = \dfrac{I_o V_o}{2} = \dfrac{(1250)(100)}{2} = 6.25 \times 10^{4} W$ 0

(iii) $\dfrac{6.25 \times 10^{4} J}{8} \times \dfrac{1 \times 10^{-3} s}{1ms} \times 0.40\, ms = 25J$ 0

(b)   0

 There can be no credit here. Although the equation is correct, it is unclear where the numbers come from.

 The physics here is difficult to understand. The average power value calculated earlier is multiplied by, among other things, time. This receives no credit.

This answer achieved 5/7

 This student remembers that the peak pd will double (Faraday's law) but forgets to show that the frequency doubles.

The negative half of the drawn curve is really not good enough. It is too pointed and messy. The positive half is much better. There is nothing to stop you turning the paper upside down to get two good halves!

This is a good answer, clearly set out, marred only by the omission of units. The examiner penalizes this the first time, but not the second.

This is a very clear solution. The student has shown on the graph where the reading was taken.

 There is no need to carry the units through the calculations, as has been done here. It is confusing and may lead you into errors by perhaps substituting the value for *V* too many times.

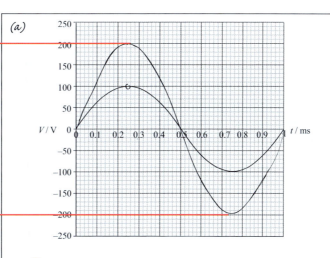

(a)

(i) $R = \dfrac{V_o}{I_o}$ $I_{rms} = \dfrac{I_o}{\sqrt{2}}$

$I_o = \dfrac{V_o}{R}$ $= \dfrac{4A}{\sqrt{2}}$

$= 100V/25\,\Omega$ $= 2.8A$

$= 41$ 1

(ii) $P_{av} = \tfrac{1}{2}\,I_o V_o$

$= \tfrac{1}{2}\,(4A)(100V)$

$= 200W$ 1

(iii) $P = V^2/R$

$= (60V)^2/25\,\Omega$

$= 144W$ 2

(b)   1

This answer achieved 7/7

This is a neatly drawn and correct graph. There is no need to take the curve through to the right-hand side providing all the physics points are there in one cycle.

Two well-set-out solutions are given.

✓ Here, as earlier, the student is taking a risk. Every final answer is given to 1 sf. Here there is always an earlier answer that goes to more sf and the student has made a conscious decision to round to a very low number of places. This is technically incorrect. Always quote your answers to the same number of sf as the question or to one more.

✓ The student shows why the frequency doubles using an argument involving the time period. This is not necessary. The information is on the graph and this piece of work serves only to waste time. If the examiner required this, the question would say: "**Sketch, explaining your answer …**" or something similar.

(a)

(i) $R = \dfrac{V_o}{I_o}$ $RI_o = V_o$ $I_o = \dfrac{V_o}{R} = \dfrac{100V}{25\,\Omega} = 4A$

$I_{rms} = \dfrac{I_o}{\sqrt{2}} = \dfrac{4A}{\sqrt{2}} = 2.8284A \Rightarrow 2.8A \Rightarrow 3A$ 0

(ii) $P_{av} = \frac{1}{2} I_o V_o$

$= \frac{1}{2}(4A)(100V)$

$= 200W$

$= 2 \times 10^2\,W$ 3

(iii) at 0.40 ms V = 60V

$P = \dfrac{V^2}{R} = \dfrac{(60V)^2}{25\,\Omega} = 144W$

$\Rightarrow 1 \times 10^2\,W$ 2

(b) $T_{original} = 1.00\,ms$

$f_{original} = \dfrac{1}{T_{orginal}} = \dfrac{1}{1.00\,ms} = 1.00\,Hz$

$\therefore 2f = 2.00\,Hz$

$T_{new} = \dfrac{1}{2.00\,Hz} = 0.500\,s$ 2

12. Atomic and nuclear physics, and quantum physics

Key terms for this chapter

- isotope—two forms of an element that are chemically identical but differ in mass
- nuclide—a nucleus expressed in its symbol form with nucleon and proton number
- nucleon—a particle that, in different charge states, is a proton or a neutron
- proton number, neutron number and nucleon number
- radioactive half-life—time taken for activity to halve or original number of atoms to halve
- decay constant—probability of decay of a nucleus per unit time
- artificial transmutation—induced change of a nucleus into a different element
- unified atomic mass unit—this is 1/12th the mass of an isolated atom of carbon-12
- mass defect—difference between measured mass and total mass of all nucleons in a nucleus
- binding energy, binding energy per nucleon—energy required to disassemble a nucleus into its neutrons and protons
- fission and fusion
- photoelectric effect—emission of electrons from metal surface when electromagnetic radiation (photons) incident on the surface
- de Broglie hypothesis
- Schrödinger model
- Heisenberg principle

The atom

You should know:

- the evidence that supports the nuclear model of the atom
- the evidence for the existence of atomic energy levels—emission and absorption spectra
- the terms nuclide, isotope and nucleon
- the definition of nucleon number, proton number and neutron number.

You should be able to:

- describe a simple model of the atom—a small nucleus surrounded by electrons, held together by electrostatic forces
- outline a limitation of this simple model of the nuclear atom
- describe the interactions that occur in the nucleus.

Example

The atomic line spectra of elements provides evidence for the existence of

A. photons.
B. electrons.
C. quantized energy states within nuclei.
D. quantized energy states within atoms.

Answer: D

The line spectra arise when electrons move from a higher energy level to a lower level. In order to lose their energy, they must emit a photon (electromagnetic radiation). The electrons undergoing this energy change are outside the nucleus. However, the spectra themselves do not give evidence for the electrons or for the photons.

Be prepared

- One of the strongest pieces of evidence for the nuclear model is that of Geiger and Marsden. They bombarded gold nuclei with alpha particles. The scattering that resulted gave Rutherford clues as to the nature of the atom. You should be familiar with the details of this experiment and also with a simple treatment of the energy changes when the alpha particle approaches the nucleus.

The atom (continued)

- You should understand the nature of the photon as a "packet of energy". This is in apparent contradiction to the view of light as a wave.

- Be aware of the nature of the Coulomb interaction between charged particles (in the nucleus, between charged particles) and the strong nuclear interaction

(between all nucleons). You need to understand the role of these two forces. Gravitational forces are exceptionally weak in comparison with the Coulomb and the other nuclear forces.

- Know the definitions of, and the notation for, the proton and nucleon numbers of nuclides.

Radioactive decay

You should know:
- that radioactive decay is random and spontaneous
- the definition of radioactive half-life
- why some nuclei are stable but others are unstable
- the biological effects of ionizing radiation
- the radioactive decay law
- the definition of decay constant
- methods for measuring the half-lives of isotopes.

You should be able to:
- describe the phenomenon of natural radioactive decay
- describe the properties of alpha (α) and beta (β) particles and gamma (γ) radiation
- determine the half-life of a nuclide from a decay curve
- describe β⁺ decay
- derive the relationship between decay constant and half-life
- solve problems involving radioactive half-life that use integral values of half-life
- solve problems involving radioactive half-life using decay constants.

Example

The decay constant of a radioactive isotope is $0.02\,\text{s}^{-1}$. Which of the following statements about this isotope is true?

A. The half-life of the isotope is $\dfrac{1}{0.02}\,\text{s}$.

B. In 1 s, 0.02 nuclei will decay.

C. A nucleus decays in every 0.02 s.

D. The probability that a nucleus will decay in 1 s is 0.02.

Answer: D

The decay constant of an isotope is the probability that any one nucleus will decay in unit time (in one second). That is the key to this question. If you really understand this, then one of the key equations in radioactive decay will be clear: the number of nuclei in a sample decaying in one second is equal to the decay constant multiplied by the total number of nuclei present.

Be prepared

- You should be aware of the mechanics of the emission of α, β⁻ and β⁺ particles, and the prediction and discovery of the neutrino. Although these emissions are nuclear events, they still have to obey the laws of momentum and energy conservation. These laws explain why the energies of the emitted particles differ in a fundamental way and why the neutrino is involved.

- Exam questions about half-life determination can be asked in a number of ways.
  - From a decay curve (it is advisable to use at least three changes and calculate their average)
  - Involving integral numbers of half-lives (each half-life reduces the number of parent nuclei by half)
  - Involving a calculation based on $N = N_0\,e^{-\lambda t}$

- Methods for determining half-lives of isotopes can involve isotopes that:
  - are very long-lived (measure the decay rate, know the number of atoms in the sample and compute λ, hence half-life)
  - have short half-lives of the order of minutes or hours (plot a graph and use it to determine the half-life)
  - have short half-lives of the order of seconds (use an ionization chamber and measure the current, leading to a determination of λ).

Nuclear reactions, fission and fusion

You should know:

- what is meant by artificial (induced) transformation
- the definitions of:
  - unified atomic mass unit
  - mass defect
  - binding energy
  - binding energy per nucleon
- what is meant by nuclear fission
- what is meant by nuclear fusion
- that the main source of the Sun's energy is nuclear fusion.

You should be able to:

- construct and complete nuclear equations
- apply the Einstein mass–energy relationship
- draw and annotate the graph that shows the variation with nucleon number of the binding energy per nucleon, and be able to apply it to explain why energy is released in fusion and fission processes
- solve problems involving mass defect and binding energy
- solve problems involving fission and fusion reactions.

Example

Two light nuclei of masses m_1 and m_2 fuse in a nuclear reaction to form a nucleus of mass M. Which of the following expressions correctly relates the masses of the nuclei?

A. $M > m_1 + m_2$

B. $M < m_1 + m_2$

C. $M = m_1 + m_2$

D. $M = m_1 - m_2$

Answer: B

This question focuses on the origin of the energy in nuclear fusion and fission. In both processes, mass m is converted into energy E according to Einstein's equation $E = \Delta mc^2$ (c is $3 \times 10^8\ m\,s^{-1}$ and Δ means "change in"). The question asks about fusion, in which two light nuclei fuse to form a more massive one. What does the equation tell you about the mass of the final nucleus?

Be prepared

- Nuclear fission is not the same as radioactive decay—it is a common error to confuse the two. Fission is always triggered by the arrival of a neutron. Radioactive decay happens spontaneously, and the probability of it happening to a particular nuclide is fixed.

- Do not confuse the words "fission" and "fusion" in an exam.

- You can be asked to write about the meaning of the graph of binding energy per nucleon against nucleon number or to use it in a description or calculation. There are three regions of the graph to which you should pay particular attention.

Binding energy per nucleon/MeV

- The region with low nucleon number, which includes elements such as hydrogen and helium. Because a helium ($^4_2$He) nucleus is especially stable, there is an energy release when it is built up from four hydrogen nuclei.
- The region around the element iron. This is the area of the greatest stability.
- The region from iron to the heaviest nuclei. Fission reactions obtain their energy by splitting the uranium (or plutonium) nucleus into two lighter nuclei, and in doing so the binding energies per nucleon move towards the iron region. This means a greater stability, so energy has to be released and the total mass is reduced.

- Try to get plenty of practice in carrying out calculations of binding energy. There are two pitfalls here, the units and the significant figures. You can meet a number of units in this type of calculation: unified atomic mass units (amu, abbreviated u—only used in atomic and nuclear physics), kg and MeV c^{-2} (which are found mostly in option H). Concentrate on the first two, but take care with conversions. Carry the significant figures through on your calculator as long as possible, making the final decision about the appropriate number at the end of the calculation.

Quantum mechanics

You should know:

- what is meant by the photoelectric effect
- the concept of the photon
- what is meant by the de Broglie hypothesis
- the concept of matter waves
- a laboratory procedure for producing and observing atomic emission and absorption spectra
- how atomic spectra provide evidence for the quantization of energy in atoms.

You should be able to:

- use the concept of the photon to explain the photoelectric effect
- describe and explain an experiment that tests the Einstein model of photoelectricity
- solve problems involving the photoelectric effect
- outline the Davisson–Germer experiment to verify the de Broglie hypothesis
- solve problems involving matter waves
- calculate wavelengths of spectral lines from energy level differences
- explain the origin of atomic energy levels in terms of the "electron in a box" model
- outline the Schrödinger model of the hydrogen atom
- outline the Heisenberg uncertainty principle in terms of both position–momentum and time–energy.

Example

In an experiment to investigate the photoelectric effect, monochromatic light is incident on a metal surface. The photoelectric current and the maximum kinetic energy of the photoelectrons are measured.

Which one of the following correctly shows the change, if any, in the photoelectric current and in the maximum kinetic energy of the photoelectrons when light of the same intensity but higher frequency is incident on the same metal surface?

	Photoelectric current	Maximum kinetic energy
A.	decreases	no change
B.	decreases	increases
C.	no change	decreases
D.	no change	increases

Answer: B

Photoelectric questions need some care. Always think about photons and never about waves in this part of the syllabus. You are asked to think about light of the same intensity but higher frequency. What is the consequence of this change? Remember that E = hf, so for each photon a higher frequency gives a higher energy. So, if the intensity is the same, the number of photons arriving per second must have gone down. What does this mean for the photoelectric current? Each photon has more energy to give to the released electron, but the electron requires a fixed amount of energy to be released from the surface. What happens to the remainder?

Be prepared

- A photon is a "packet" of electromagnetic energy. Some physics phenomena cannot be explained in terms of light as a wave—photoelectricity is one of them. An electron needs energy if it is to leave the surface of a metal. If a photon arrives at the metal surface with enough energy, then the electron and photon can interact and the electron is released. Any excess energy that the photon brings with it is given to the photoelectron as kinetic energy.

- If the irradiated metal is charged, then this affects the outcome. A metal that has a negative charge will repel the electrons when they are released. A positive charge on the metal will attract the electrons back into the surface and the electron may not be detected.

- Einstein's explanation of photoelectricity included the photoelectric equation. Make sure that you can identify the three terms in the equation and describe what they represent (the incoming energy of the photon, the work function of the metal, and the surplus kinetic energy of the released electron).

- You may well have been shown or have carried out an experiment to test the Einstein model. Make sure that you understand both the method and the explanation for all the effects you saw.

- De Broglie suggested the idea of matter waves because, he argued, if light can act like a particle, then mass ought to have wave properties.

- You may be asked to calculate the wavelength of moving charged particles from knowledge of the pd through which a charged particle has been accelerated. Or you may be given the kinetic energy—remember that $\frac{1}{2}mv^2$ can also be written as

$$\frac{1}{2}\frac{(mv)^2}{m}$$

Practise manipulating these equations.

Quantum mechanics (continued)

- An emission spectrum forms from the light emitted when electrons fall to a lower energy level.

- An absorption spectrum is observed when radiation passes through a material and electrons are promoted to a higher energy. As a result, energy is removed from the radiation, leaving (typically) a dark line in the spectrum.

- Monatomic, low-density gases have emission spectra that are series of lines. Each line corresponds to one distinct energy change. It is possible to deduce the energy structure of the atomic electrons from the spectrum of an element.

- Calculations will again be based on $E = hf$, but, as with other parts of this topic, understanding the units is key. You could be given energies in joules, electronvolts or multiples of these, and you need to be completely confident about converting from one unit to another.

- The "electron in a box" model due to Schrödinger suggests that the electron associated with a nucleus can be described as a probability wave. There are only certain probability waves that allow the electron to be trapped in the atom—others would allow it to escape. So any atoms that exist must have electrons with these special "bound" probability wave shapes. It turns out that the mathematics of this predicts behaviour where the energy of the electron is related to hf, suggesting that quantization occurs. This corresponds exactly to the idea of matter waves.

- Heisenberg took a separate view of particles, and said that it was not possible to say exactly where a particle is located and what its momentum is at exactly the same moment. The argument also applies to the energy of an object and the time at which the energy is measured. These are difficult ideas, but you can think of uncertainty in terms of the blurring out of the exact position of the electron by the probability wave that Schrödinger describes.

Nuclear physics

You should know:

- how nuclear radii may be estimated from charged-particle scattering experiments

- how a Bainbridge mass spectrometer can be used to determine the masses of nuclei.

You should be able to:

- describe one piece of evidence for the existence of nuclear energy levels.

Example

The diagram is a schematic representation of the Bainbridge mass spectrometer. Positive ions are injected between the plates of the speed selector.

Which of the following correctly shows the direction of the magnetic fields B_1 and B_2?

	B_1	B_2
A.	out of the page	out of the page
B.	into the page	into the page
C.	out of the page	into the page
D.	into the page	out of the page

Answer: A

The Bainbridge mass spectrometer is an instrument that can be used to determine the masses of atoms. The region B_1 is used as a "velocity selector" so that all ions going through both slits have the same speed as each other. This is achieved by having two fields, one electric (shown in the diagram) and one magnetic. The two fields are crossed, and for one unique speed give the same magnitudes of electrostatic and magnetic forces on the ion but acting in opposite directions. This information should allow you to work out the required direction of the B_1 field. When the ions, all travelling at the same speed, enter region B_2, there is only one field, magnetic, and the drawn curvature should allow you to work out the correct direction, knowing that a centripetal force is required.

Nuclear physics (continued)

Be prepared

- If alpha particles with a particular energy are fired head-on at a nucleus, they will, at some distance from the centre of the nucleus, stop and reverse their direction. The higher the initial kinetic energy of the alpha particle, the closer it will get to the nucleus before it reverses its path. Experiments like this with charged particles allow estimates of the size of nuclei. You can be asked to solve numerical problems on this.

- One piece of evidence for the existence of nuclear energy levels is the observed energies of alpha particles that are emitted in alpha decay. There is a link between the alpha energies and the gamma ray energies that are emitted at the same time. If a daughter nucleus ends up in the ground state after alpha emission, then no gamma rays are observed. However, if the alpha particle has less energy than the total change would suggest, then a gamma photon is emitted. This gamma radiation has exactly the right amount of energy to make up the difference. So it is thought that there are energy levels in the nucleus in a similar way to the electron energy levels of the atom.

Part 2   This question is about atomic spectra.

(a)   The diagram represents some of the energy levels of the mercury atom.

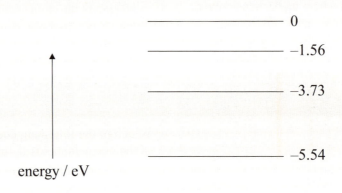

Photons are emitted by electron transitions between the levels. On the diagram draw arrows to represent the transition, for those energy levels that gives rise to,

(i)   the longest wavelength photon (label this L).   [1]

(ii)   the shortest wavelength photon (label this S).   [1]

(b)   Determine the wavelength associated with the arrow you have labelled S.   [3]

Nuclear spectra

(c)   A nucleus of the isotope bismuth-212 undergoes α-decay into a nucleus of an isotope of thallium. A γ-ray photon is also emitted.

Draw a labelled nuclear energy level diagram for this decay.   [2]

(d)   The activity of a freshly prepared sample of bismuth-212 is 2.80×10^{13} Bq.   After 80.0 minutes the activity is 1.13×10^{13} Bq. Determine the half-life of bismuth-212.   [4]

[Taken from HL paper 2, time zone 1, May 2009]

How do I approach the question?

(a) The question is asked in terms of wavelengths. Remember that $E = hf$ or $E = ch/\lambda$, where c is the speed of light in a vacuum. A long wavelength has a small frequency and a small energy. Your job is to identify the transitions and to mark them with the correct directions.

(b) You will need the equation mentioned in part (a) to calculate the wavelength. Take care with powers of 10 and the units.

(c) The alpha particle energy together with the gamma-ray photon's energy must be equal to the total energy change between the bismuth and the thallium. You will

need to draw a well-labelled diagram that makes the order of these transitions and the energy levels clear.

(d) You will need to be able to use the exponential radioactive equation

$N = N_0 e^{-\lambda t}$.

You will also need to remember the relationship between λ and the half-life, as the half-life will be your final answer:

$$\lambda = \frac{\ln 2}{t_{\frac{1}{2}}}$$

in an obvious notation.

This answer achieved 0/11

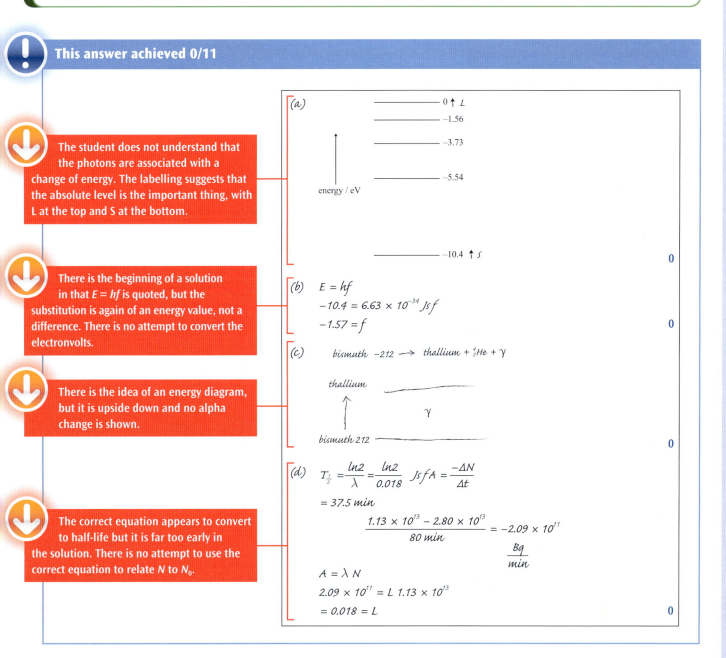

The student does not understand that the photons are associated with a change of energy. The labelling suggests that the absolute level is the important thing, with L at the top and S at the bottom.

There is the beginning of a solution in that $E = hf$ is quoted, but the substitution is again of an energy value, not a difference. There is no attempt to convert the electronvolts.

There is the idea of an energy diagram, but it is upside down and no alpha change is shown.

The correct equation appears to convert to half-life but it is far too early in the solution. There is no attempt to use the correct equation to relate N to N_0.

This answer achieved 6/11

Although the correct levels are here, the arrows are going in the wrong direction. The student thinks that the 0 eV level is the ground state.

The correct equation is used, but the conversion to joules from electronvolts is missing, and as a result the frequency value is wrong.

The final answer is of the order of 10^{42} m. As a distance, it is much greater than the size of the universe. The student should have noticed how improbable this was. Answers in IB exams will never be implausible.

There is credit for converting the incorrect frequency to a wavelength.

No energy differences are indicated. The diagram is probably upside down.

This could have gone badly wrong! The notation in the equation is unusual and the power should be negative (this is corrected later).

Although this is given full credit, the calculation of the half-life from λ is long-winded. It would have been better simply to substitute into the equation from the *Physics data booklet* rather than go through this further substitution into $N = N_0 e^{-\lambda t}$.

(a)

energy / eV

0
L —1.56
—3.73
—5.54
s
—10.4

1

(b)

$E = hf$

$E = 10.4 eV$

$h = 6.63 \times 10^{-39}$ Js

$f = ?$

$10.4 eV = 6.63 \times 10^{-34}$ Js (f)

$f = 1.56862745 \times 10^{-34}$ s^{-1}

$V = f \cdot \lambda \qquad v = c$

$\lambda = \dfrac{v}{f} = \dfrac{c}{f}$

$\lambda = \dfrac{3.00 \times 10^8 \text{ m/s}}{1.56862745 \times 10^{-34} \text{ s}^{-1}}$

$\lambda = 1.91 \times 10^{42}$ m

1

(c)

energy/ev

γ ———

α ———

0

(d) $y = Ce^{kt} \quad t = 0, y = 2.80 \times 10^{13}$ Bq

$2.80 \times 10^{13} = C$

$Y = (2.80 \times 10^{13})e^{kt} \quad t = 80.0, y = 1.13 \times 10^{13}$ Bq

$1.13 \times 10^{13} = (2.80 \times 10^{13})e^{k(80.0 min)}$

$\ln[.404 = e^{k(80.0 min)}]$

$-.907 = k (80.0 min)$

$K = -.0113$ min^{-1}

$y = 1.4 \times 10^{-13}, t = ?$

$y = (2.80 \times 10^{13})e^{(-.0113)t}$

$\ln[.5 = e^{(-.0113)t}]$

$-.693 = (-.0113)t$

$t = 61.1$ min   half life

4

This answer achieved 9/11

The transitions are clear and in the correct direction.

The working is clear and it is obvious what the student is doing. The answer is to an appropriate number of significant figures.

The student has not read the question carefully and thinks that a nuclear equation is required.

This is a clear solution, slightly unusual in that the student calculates N/N_0 rather than substitating time into the equation.

The working is much clearer than previous examples. Try to set your work out in this logical way.

(a)

```
                          0
     ↓ L            ──── −1.56
                    ──── −3.73

                    ──── −5.54
  energy / eV

                S

                    ──── −10.4
```
2

(b) $E = hf = \dfrac{hc}{\lambda}$ $10.4\,eV = \dfrac{hc}{\lambda}$ $\lambda = \dfrac{hc}{10.4eV}$

$\lambda = \dfrac{hc}{1.66 \times 10^{-18}\,J}$ $= 1.20 \times 10^{-7}\,m$ 3

(c) $^{212}Bismuth \longrightarrow \,^{208}Thallium* + \,^{4}_{2}\alpha$

$^{208}Thallium* \longrightarrow \,^{208}Thallium + \,^{0}_{0}\gamma$ 0

(d) $A/A_0 = 1.13 \times 10^{13}/2.80 \times 10^{13} = 0.404$

in 80 min activity has decreased by a factor activity $\propto N$, so;

$0.404\,N_0 = N_0 e^{-\gamma(4800s)}$

$0.404 = e^{-\gamma(4800s)}$

$-0.906 = -\lambda(4800\,s)$

$\lambda = 1.89 \times 10^{-4}\,s^{-1}$

$T_{\frac{1}{2}} = \dfrac{ln2}{\lambda} \Rightarrow$ $T_{\frac{1}{2}} = \dfrac{ln2}{1.89} \times 10^{-4}s^{-1}$

$T_{\frac{1}{2}} = 3670\,s$ 4

Part 2   This question is about the decay of radium-226.

(a)   A nucleus of the isotope radium-226 (Ra) undergoes α-decay with a half life of 1.6×10^3 yr to form a nucleus of radon (Rn).

Define the terms *isotope* and *half-life*.   [2]

Isotope:

Half-life:

(b)   Using the grid below, sketch a graph to show how the activity A of a sample of radium-226 (Ra) would be expected to vary with time t over a period of about 5.0×10^3 yr . The activity of the sample at time $t = 0$ is A_0.   [3]

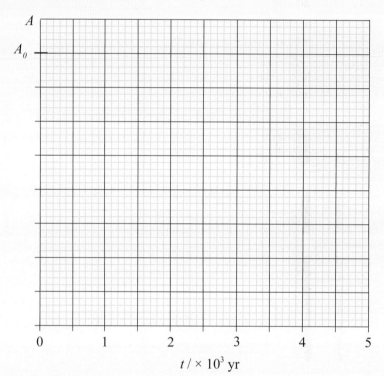

$t / \times 10^3$ yr

(c)   The nuclear reaction equation for the decay of radium-226 (Ra) may be written as

$$^{226}_{88}\text{Ra} = \text{Rn} + \alpha$$

(i)   State the value of the proton number and neutron number of the isotope of radon (Rn).   [1]

Proton number:

Neutron number:

(ii)   Outline why the binding energy of Ra is less than that of Rn.   [2]

(d)   The following data are available.

mass of Ra   = 226.0254 u
mass of Rn   = 222.0175 u
mass of α = 4.0026 u

Show that the energy released in the decay of a Ra nucleus is 4.94 MeV.   [2]

[Taken from SL paper 2, time zone 1, May 2009]

How do I approach the question?

(a) Check that you know the definitions before beginning your answer.

(b) You need to draw a smooth curve showing half-life behaviour.

(c) (i) The solution begins with your knowledge of the constituents of an alpha particle and what the nucleon and mass numbers are for it. From these, you can deduce the relevant numbers for the radon.

    (ii) For a nuclear decay to proceed, energy must be given out in the reaction. The energy change leads to changes in the binding energy per nucleon (radon is higher up the slope of the binding energy per nucleon graph compared to radium).

(d) You will need to find the conversion of atomic mass units (u) into kilograms. Calculate the mass change in the reaction and from there use Einstein's energy–mass equation to arrive at an energy value in joules. One last conversion to energy in MeV should give the answer. As it is a "**show that**" question, give all steps of the calculation in a logical sequence.

Which areas of the syllabus is this question taken from?

- This question requires knowledge of "The atom" (topic 7.1) and "Nuclear reactions, fission and fusion" (topic 7.3).

This answer achieved 2/10

There are many stable isotopes of elements.

The problem here is the use of the words "decay by 50%". The answer needs to make it clear that, after one half-life, only one-half of the original substance remains. Alternatively, it is the time after which the rate of decay has halved.

The curve is not drawn carefully enough, it is not smooth and it is not one continuous curve.

There is a recognition that 222 is important because the nucleon number of the radon is 222 and the alpha nucleon number is 4. But both parts need to be correct to gain the 1 mark.

This comment has no physical basis.

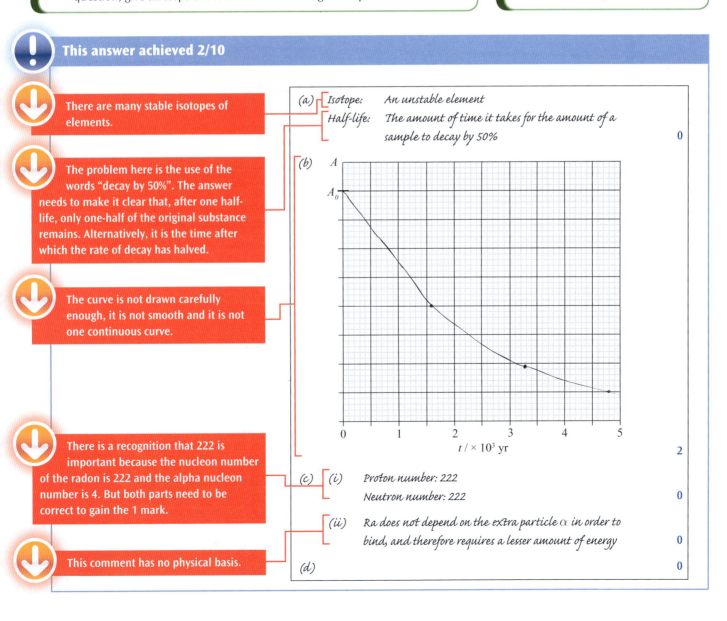

(a) Isotope: An unstable element

    Half-life: The amount of time it takes for the amount of a sample to decay by 50%   **0**

(b)   **2**

(c) (i) Proton number: 222

    Neutron number: 222   **0**

    (ii) Ra does not depend on the extra particle α in order to bind, and therefore requires a lesser amount of energy   **0**

(d)   **0**

This answer achieved 7/10

Two good definitions are given.

Compare this half-life definition with the previous one. The idea of decaying into another element is crucial.

This curve is well drawn and goes through all the points. Possibly the student's original line was a little thick.

There is only 1 mark for the whole of part (i), so the failure in the neutron number disqualifies everything.

The idea that radium is less stable is given credit.

What is missing is the idea that, because the binding energy per nucleon is less in radium, therefore in changing to radon there is more binding energy per nucleon, and this difference in binding energy per nucleon has to be lost by the nucleus.

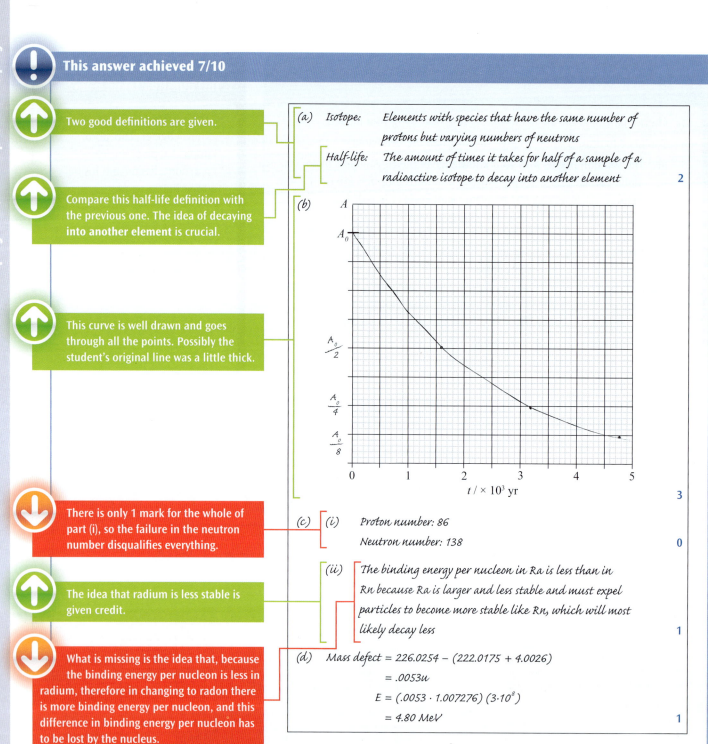

(a) Isotope: Elements with species that have the same number of protons but varying numbers of neutrons

Half-life: The amount of times it takes for half of a sample of a radioactive isotope to decay into another element

2

(b)

3

(c) (i)   Proton number: 86
Neutron number: 138

0

(ii) The binding energy per nucleon in Ra is less than in Rn because Ra is larger and less stable and must expel particles to become more stable like Rn, which will most likely decay less

1

(d)   Mass defect $= 226.0254 - (222.0175 + 4.0026)$
$= .0053u$
$E = (.0053 \cdot 1.007276)(3 \cdot 10^8)$
$= 4.80 \, MeV$

1

This answer achieved 8/10

These definitions are too vague and incomplete for credit.

(a) Isotope: A species of an element with a specific mass number.
Half-life: The amount of time required for half of a sample to decay.

0

(b)

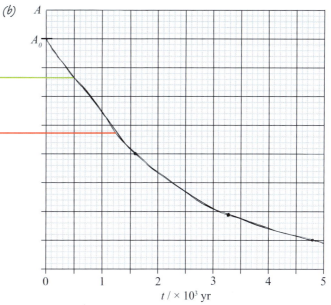

$t / \times 10^3$ yr

3

The shape of the graph is good.

The line is kinked and double in places. This has not been penalized here, but would probably be penalized in the data-analysis question in paper 2.

This is correct (obviously after an initial re-think). Do not be afraid to change your mind, but make corrections as clear as possible.

(c) (i) Proton number: 86
Neutron number: ~~222~~ 136

1

This is awarded 1 mark for the mention of stability, and, with the benefit of the doubt, the statement about breaking up the nucleus is correct too.

(ii) Nuclei always decay into more stable forms, so Ra is less stable than Rn. This means that less energy per nucleon is required to break up the nucleus of Ra.

2

This solution is worked in different units from the previous one. It uses atomic mass units and the direct conversion to MeV c^{-2} that is given in the *Physics data booklet*.

(d) $\Delta m = 226.0254\ u - 222.0175\ u - 4.0026\ u$
$\Delta m = .0053\ u$
$\Delta E = .0053\ uc^2 \dfrac{931.5\ MeV\ c^{-2}}{u} = 4.94\ MeV$

2

13. Energy, power and climate change

Key terms for this chapter

- Sankey diagram
- energy density—useful energy stored per unit volume or per unit mass depending on the fuel
- renewable and non-renewable resources
- controlled and uncontrolled nuclear fission, and nuclear fusion
- chain reaction, fuel enrichment, moderator, control rod, heat exchanger and neutron capture
- solar heating panel, photovoltaic cell
- lake-water storage, tidal-water storage, pump storage
- wind generator
- oscillating water-column wave generator
- albedo, black-body radiation and Stefan–Boltzmann law
- emissivity
- coefficient of volume expansion

Energy degradation and power generation

You should know:

- what is meant by degraded energy
- that the continuous conversion of energy into work requires a cyclic process and the transfer of some energy from the system.

You should be able to:

- construct and interpret Sankey diagrams
- outline the mechanisms involved in the production of electrical power.

Example

Degraded energy is energy that

A. has not yet been used.
B. cannot be used further.
C. is always electrical energy.
D. is always potential energy.

Answer: B

Energy is transferred from one form to another in commercial power generation. The generator will always dissipate some energy that is no longer useful. A good example of this is the friction at the bearings in an electrical generator. The energy eventually lost to heat in this case is an example of degraded energy. It is difficult to use it further. On the other hand, electrical energy is in a high-grade form. Potential energy (the question does not say which type) is normally the starting point for an energy conversion not the end point.

Be prepared

- In a Sankey diagram, the width of the line indicates the relative size of an energy contribution. Use the diagram with care and either measure the widths carefully or read off the width from the grid on the paper. It is a convention that the energy contributions appear on the diagram in their correct order in the overall process. Similarly, if you have to draw a Sankey diagram, make sure that the widths are accurate.
- Electrical power is often generated from fuels that are burnt to heat water to form steam. This steam is used to power a turbine, which is attached to a dynamo, and it is the dynamo that actually converts the energy of the steam into an electrical form. Do not forget the step from turbine to dynamo.

World energy sources

You should know:

- the definition of energy density
- the relative proportions of world use of the different energy sources available.

You should be able to:

- discuss how the choice of a fuel is influenced by its energy density
- distinguish between renewable and non-renewable resources
- discuss the relative advantages and disadvantages of various energy sources.

Example

The unit of energy density of a fuel is

A. $J\,m^{-2}$.

B. $J\,m^{-1}$.

C. $J\,kg^{-1}$.

D. $kg\,J^{-1}$.

Answer: C

There are a number of cases in physics where the term "density" is used in a context where it does not have its usual meaning of "per unit volume". This can be one of them. One way to define the energy density of a fuel is as a measure of the number of joules of energy that can be released from unit mass of the fuel. So the unit will be related to mass not volume in this case.

Be prepared

- Fuels will be chosen for a particular situation by, among other things, their energy density. Consider the different needs of a spacecraft on a deep-space mission and a steam train that can refuel with coal when required.

- Do not be confused by the difference between non-renewable and renewable resources. Although we do not consider non-renewables such as coal or oil to be re-forming in our own lifetimes, nevertheless, there is a slow production process going on. The difference between non-renewable and renewable is in the **rate** of formation. Renewable forms regenerate rapidly. The rate of formation of a non-renewable is less (usually much less) than the rate of consumption.

- Be careful to use clear and accurate statements of physics in discussing resources. It is, for example, wrong to suggest that "renewable resources can be used again". The examiner will not give credit for this, because a particular resource cannot be used more than once. Equally, do not suggest that non-renewables can be re-formed in "thousands of years"—the timescale is much longer than this.

- You should memorize at least one advantage and one disadvantage for each of the major fuel groups.

Fossil fuel power production

You should know:

- the historical and geographical reasons for the widespread use of fossil fuels
- the relative advantages and disadvantages associated with the transportation and storage of fossil fuels
- the approximate overall efficiency of power stations fuelled by different fossil fuels
- the environmental problems associated with the recovery of fossil fuels and their use in power stations.

You should be able to:

- estimate the rate of fuel consumption by power stations, knowing the energy density of the fuel.

Example

Which of the following is the best estimate for the overall efficiency of a typical coal power station?

A. 5%

B. 30%

C. 60%

D. 90%

Answer: B

You need to know the approximate efficiencies of fossil fuel power stations. This question focuses on a coal-fired station. The efficiencies for these vary from around 30% to 40%. Gas-fired stations are a little higher, ranging from 30% to 45%, but

Fossil fuel power production (continued)

the efficiencies of both types are gradually improving all the time. Combined heat and power plants are able to give the best efficiencies, sometimes over 50%.

You are always asked to give the best estimate or the closest answer in all multiple-choice questions for the IB, so in this type of question the responses will be well spread out and you have to give the most appropriate answer.

Be prepared

- You may be asked to carry out a multi-step calculation to determine the consumption of a fuel by a power station. This involves taking the energy-density figures for a fuel and the required power output of the station to calculate the mass of fuel needed every second. If you are asked for the consumption per hour or per day, do not forget the conversion from seconds.

Non-fossil fuel power production—nuclear power

You should know:

- how neutrons from a fission reaction can initiate further fissions (a chain reaction)
- the difference between controlled and uncontrolled nuclear fission
- what is meant by fuel enrichment
- the main energy transformations that take place in a nuclear power station
- what the roles of the following are in a nuclear power station:
  - moderator
  - control rod
  - heat exchanger
- how neutron capture leads to the production of plutonium
- the problems associated with nuclear fusion.

You should be able to:

- describe the importance of plutonium
- discuss the safety issues associated with nuclear power
- solve problems on the production of nuclear power (fission and fusion).

Example

In a nuclear power station, uranium is used as the energy source and plutonium-239 is produced. Which of the following is true?

A. Plutonium-239 is produced by nuclear fusion.

B. A moderator is used to absorb plutonium-239.

C. Control rods are used to slow down plutonium-239.

D. Plutonium-239 can be used as a fuel in another type of nuclear reactor.

Answer: D

This question tests your understanding of a number of different topics within nuclear power. You need to understand the roles of the moderator and the control rods in the fission reactor, and you need to know the types of fuel and the products that are found in the reactor. Finally, the question checks that you know not only that there are uranium-based fission reactors, but also that there are fast-breeder reactors, which use a different fuel that is itself produced in conventional uranium-based reactors.

Be prepared

- Questions on this topic may not necessarily be set in an energy-resource context. They can arise within a question on topic 7 (nuclear physics), as the two separate parts of the syllabus are closely related. Make sure that you have linked your understanding of both parts of the syllabus.
- Know all the energy transformations that take place in a nuclear power station and ensure that you can both describe the processes and draw diagrams to represent them.
- Link all the parts in a question. If a question draws specifically on a particular nuclear fission reaction in which (say) three neutrons are emitted, then in any further discussion of the reaction—perhaps asking about a chain reaction—make sure that you draw or describe the reaction in terms of the three neutrons.

Non-fossil fuel power production—solar power

You should know:

- the difference between a solar heating panel and a photovoltaic cell

- the reasons for seasonal and regional variations in the solar power incident on a square metre of Earth's surface.

You should be able to:

- solve problems involving applications of photovoltaic cells and solar heating panels.

Example

When sunlight is incident on a solar cell, an electric current is produced. This is due to

A. a temperature gradient within the cell.

B. very long-wavelength infrared radiation.

C. very short-wavelength ultraviolet radiation.

D. the photoelectric effect.

Answer: D

The term "solar cell" needs to be treated with care. Does the question mean "photovoltaic cell" or "solar heating panel"? The clue is in the first sentence: "an electric current is produced". So the question is talking about a photovoltaic cell. This means that temperature gradients are not what make the cell operate. Use your knowledge of Earth's atmosphere to decide whether the radiations quoted penetrate to the surface or not.

Be prepared

- Solar heating panels and photovoltaic cells are sometimes grouped together using the term "solar cells". Make sure you know which type you are writing about.

- Calculations involving these panels and cells can test your knowledge of a number of areas of the syllabus. Both types can involve the energy delivered to Earth's surface from the Sun, efficiency issues and Sankey diagrams. Photovoltaic cell calculations can involve standard electrical theory (chapter 10). Solar heating panels can involve heat-capacity calculations (chapter 8).

Non-fossil fuel power production—water and wind power

You should know:

- that hydroelectric power schemes can be based on:
  - lake water storage
  - tidal water storage
  - pump storage
- the basic features of a conventional horizontal-axis wind generator
- that the complete conversion of wind kinetic energy into mechanical kinetic energy in a wind generator is impossible
- the principle of operation of an oscillating water column (OWC) ocean-wave energy converter.

You should be able to:

- describe the main energy transformations that take place in hydroelectric schemes
- determine the maximum power that can be delivered by a wind generator
- determine the maximum power per unit length of a wavefront when the profile of the wave is rectangular
- solve problems involving hydroelectric, wave and wind power.

Example

The power per unit length P of an oscillating water column (OWC) is due to the action of a surface wave of amplitude A. Which of the following correctly relates P

and A, and correctly identifies the nature of the water column?

	Relation between P and A	Nature of energy
A.	$P \propto A$	kinetic
B.	$P \propto A$	kinetic and potential
C.	$P \propto A^2$	kinetic
D.	$P \propto A^2$	kinetic and potential

Answer: D

In this design, the oscillating water column is used to push air through a turbine. The Physics data booklet contains the equation that you need, but you need to know where to look to find it. The water in the column is rising and falling, changing its speed as it does so at the expense of its height.

Be prepared

- Most of the calculations you may be asked will require a straightforward conversion of gravitational potential energy. However, if you need to convert quantities into the most appropriate units, make sure that you do so correctly.

Greenhouse effect

You should know:

- the definition of albedo

- the factors that affect the albedo of the planet

- what is meant by the greenhouse effect

- the main greenhouse gases and their sources

- what is meant by black-body radiation

- the Stefan–Boltzmann law and how to use both it and the idea of emissivity to compare emission rates from different surfaces

- the definition of surface heat capacity.

You should be able to:

- calculate the intensity of the radiation from the Sun that is incident on a planet

- explain the molecular mechanisms by which greenhouse gases absorb infrared radiation

- analyse absorption graphs to compare the relative effects of different greenhouse gases

- understand graphs of the emission spectra of black bodies radiating at different temperatures

- to use a simple energy balance climate model to solve problems on the greenhouse effect.

Example

The average temperature of the surface of the Sun is about 20 times higher than the average surface temperature of the Earth. The average power per unit area radiated by the Earth is P. Which of the following gives the average power per unit area radiated by the Sun?

A. $20P$

B. $400P$

C. $8000P$

D. $160\,000P$

Answer: D

This question involves the use of the Stefan–Boltzmann law ($P = \sigma A T^4$). You are told the ratio of the temperatures and you can use the law to calculate the ratio of powers per unit area.

Be prepared

- Try not to confuse the greenhouse effect and the enhanced greenhouse effect. This section of the syllabus deals with the physics of the effect that makes Earth habitable as we know it. Concentrate on the physics of trapping of heat in the atmosphere due to the shift of radiation wavelength, because the Earth and Sun are at very different temperatures. The extra warming that is thought to be occurring at the moment is dealt with in the next section.

Global warming

You should know:

- some of the models that have been used to explain global warming are:
  - increased solar flare activity
  - cyclical changes in Earth's orbit
  - volcanic activity
  - changes in the composition of atmospheric greenhouse gases

- what is meant by the enhanced greenhouse effect

- that the increased combustion of fossil fuels is a likely major cause of the enhanced greenhouse effect

- some of the various mechanisms that can increase the rate of global warming are:
  - reduction of ice and snow cover on Earth's surface
  - reduction of CO_2 in the sea and a consequent increase in the atmosphere

  - deforestation leading to a reduction in carbon fixation

- the definition of coefficient of volume expansion

- that climate change is an outcome of an enhanced greenhouse effect

- some solutions to reduce the enhanced greenhouse effect.

You should be able to:

- describe the evidence that links global warming to increased levels of greenhouse gases

- explain why the mean sea level may rise as a result of global warming

- solve problems related to the enhanced greenhouse effect

- discuss the international efforts that are being made to reduce the enhanced greenhouse effect.

Global warming (continued)

Example

Global warming reduces the ice and snow cover on Earth. Which of the following correctly describes the changes in albedo and rate of energy absorption by Earth?

	Albedo	Rate of energy absorption
A.	increase	increase
B.	decrease	increase
C.	increase	decrease
D.	decrease	decrease

Answer: B

Albedo is the percentage ratio of radiation reflected to incoming radiation. If less radiation is reflected, then Earth must be absorbing more of it. Snow and ice both have a high albedo, rock does not.

Be prepared

- Remember to give the physical principles in your answers in this topic. It is easy to fall into the trap of giving general and vague answers. Although many scientists agree that global warming is due to human activity, not all scientists are convinced, and there are other plausible suggestions. You must be prepared to acknowledge the science that lies behind these.

- Try to have up-to-date knowledge of international initiatives. Not all will be mentioned in the syllabus. There has, for example, been a major conference at Copenhagen since the Diploma Programme *Physics guide* was written.

- An area of major misunderstanding is that of the different effects on the sea level caused by melting ice. The effect depends on whether the ice was originally on land (such as Greenland or the Antarctic) or floating (Arctic sea ice and icebergs calved from the Antarctic).

Part 2 This question is about global warming.

(a) One reason often suggested for global warming is the enhanced greenhouse effect.

    (i) State what is meant by the enhanced greenhouse effect. *[1]*

    (ii) State **two** other possible causes of global warming. *[2]*

       1.

       2.

(b) One effect of global warming is to melt the Antarctic ice sheet. The following data are available for the Antarctic ice sheet and the Earth's oceans.

$$\begin{aligned}
\text{Area of ice sheet} &= 1.4 \times 10^7 \, \text{km}^2 \\
\text{Average thickness of ice} &= 1.5 \times 10^3 \, \text{m} \\
\text{Density of ice} &= 920 \, \text{kg m}^{-3} \\
\text{Density of water} &= 1000 \, \text{kg m}^{-3} \\
\text{Area of Earth's oceans} &= 3.8 \times 10^8 \, \text{km}^2
\end{aligned}$$

Using the data, determine the

    (i) mass of the Antarctic ice. *[2]*

    (ii) change in mean sea level if all the Antarctic ice sheet were to melt and flow into the oceans. *[3]*

(c) Outline the difference, if any, that the melting of oceanic ice sheets makes to the mean sea level of the Earth. *[2]*

[Taken from SL paper 2, November 2009]

How do I approach the question?

(a) (i) Simply state what the enhanced greenhouse effect is thought to be caused by. Remember to write about the enhanced effect, not the effect on which the planet relies.

(ii) The examiner requires two causes of global warming other than the enhanced greenhouse effect. So an answer that gives factors that produce enhancement is not appropriate. Focus on the natural causes of global warming rather than man-made causes.

(b) (i) You are given the thickness of the ice sheet, its area and its density. Calculating the total mass should be straightforward.

(ii) The mass of the ice will not change when it melts, but because the density is different, the volume will change. Calculate this and then imagine that the water is spread out over the entire area of the oceans. This will give the height change. The Antarctic ice sheet (unlike the Arctic sheet) rests on rock.

(c) Finally you are asked to look at what happens in the case of an ice sheet that does float on water. The keys here are the principle of flotation and Archimedes principle. The ice floats on the sea water. How much is out of the surface and how much below? Imagine the space in which the ice sits. How much of this volume will the melted water occupy?

Which areas of the syllabus is this question taken from?

- This question requires knowledge of "Greenhouse effect" (topic 8.5) and "Global warming" (topic 8.6).

This answer achieved 1/10

✔ The number of answer lines should give you a good idea how much to write.

⬇ The answer is an attempt to explain the mechanism of the greenhouse effect. It does not deal with the enhanced effect.

⬇ This is trying to suggest what causes the enhanced effect. "Industrial waste" is a weak answer for this suggestion.

⬇ There is a power of 10 error from misreading of m for km. The data is to 2 sf. Writing the answer to 4 sf will lead to a significant figure penalty.

(a) (i) When gas molecules, such as CO_2 gas in the atmosphere, absorb heat from the earth, that is meant to be radiated out into space, hence the earth heats up. This method is used to keep green-houses warm and at a likeable temperature   0

(ii) 1. Human activity (industrial waste, cars etc)
2. Burning of fossil fuels   0

(b) (i) Mass = $1.4 \times 10^{10} \times 1.5 \times 10^{3} \times 920$
= 1.932×10^{16} kg   1

(ii) Change in mean sea level
=   0

(c) The mean sea level would rise, as the density of ice is less than that of water, so if the ice melted, it would take up more space, causing a rise in the mean sea level of the earth   0

⬇ This is the wrong way round! If the density of the water is greater than the density of ice, then the water will have a smaller volume than the same mass of ice.

This answer achieved 4/10

This goes beyond a statement.

Ozone depletion has little role to play in global warming. It has more influence on the amount of ultraviolet radiation reaching the surface.

This is not clear. The albedo of clean snow is constant. It is an effect of, not a cause of, global warming.

The solution uses metres instead of kilometres.

The student forgets to indicate whether the level goes up or down.

There is no clear answer. The student ignores the physics of the situation and thinks that the value depends on deviations from the average sea level.

This answer achieved 9/10

This answer makes clear what "enhanced" means and what causes the enhancement.

Two possible causes are clearly stated. Another possible answer could be volcanic activity.

Only 1 mark is awarded because some of the data are in kilometres and the student is expecting metres.

This is a good example of an error carried forward. The answer is still correct because the m/km error self-cancels. You might not score marks with working such as this in a "show that" question, however.

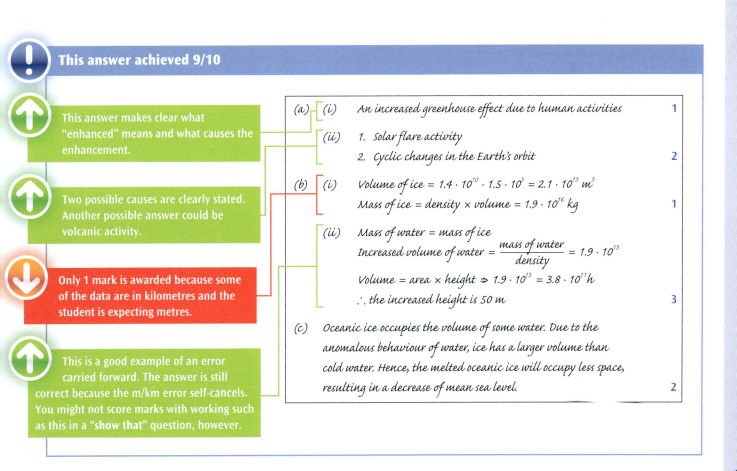

Part 2   This question is about fossil fuels.

(a)   State **two** examples of fossil fuels.   *[2]*

(b)   Explain why fossil fuels are said to be non-renewable.   *[2]*

(c)   A Sankey diagram for the generation of electrical energy using fossil fuel as the primary energy source as shown.

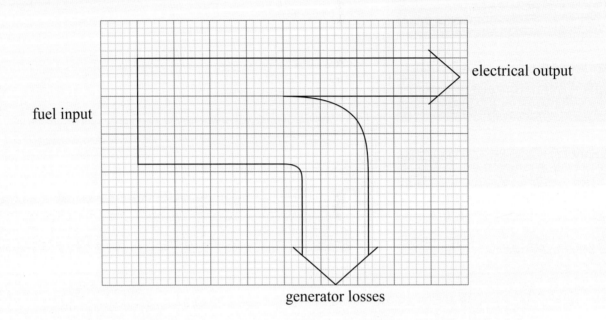

Use the Sankey diagram to estimate the efficiency of production of electrical energy and explain your answer.   *[2]*

(d)   Despite the fact that fossil fuels are non-renewable and contribute to atmospheric pollution there is widespread use of such fuels. Suggest three reasons for this widespread use.   *[3]*

1.

2.

3.

[Taken from HL paper 2, time zone 2, May 2009]

How do I approach the question?

(a) Name two examples of fossil fuels. Ensure that you are not quoting the same fuel twice by giving brand names.

(b) Bear in mind some of the advice given earlier about the physical meaning of a non-renewable fuel. Take care not to use phrases that are too simplistic.

(c) The quality of explanation is important. The diagram is given on a grid, and this indicates clearly that you are expected to take precise readings from the Sankey diagram.

(d) Ensure that you give three completely separate reasons for the uses of fossil fuels. Try to give your answers at a high level using as much physics as possible.

Which areas of the syllabus is this question taken from?

• This question requires knowledge of "Energy degradation and power generation" (topic 8.1), "World energy sources" (topic 8.2) and "Fossil fuel power production" (topic 8.3).

This answer achieved 3/9

Oil and coal are good options to choose. As this is a "state" question, do not waste time by giving any other details.

The reference to wind is good, but ambiguous, as the same wind energy certainly cannot be used twice and has to be regenerated by the Sun. The suggestion that Earth will run out of them is not worth credit.

There are lots of things wrong here. There is no explanation and it is impossible to tell where the numbers come from. There is also a significant figure penalty, as it is not possible to evaluate the answer to 4 sf from this grid.

In this type of question it is worth marking the grid to show how you are using the diagram. If you make a slip in reading from the grid, and the examiner can see your mistake, you may get some credit for later parts of the section.

(a) 1. Oil
2. Coal   2

(b) Once used they can not be reused like wind. Eventually we will run out of them   0

(c) The efficiency is $\frac{1}{3}$ = 33.33%   0

(d) 1. Fossil fuels are easy to use to generate energy
2. It is relatively easy to get coal and oil and there are huge quantities
3. It is relatively cheap and economically reliable to use fossil fuel engines   1

The separate points written here are an "ease of use" comment and an "economic" comment. These are as vague as comments can be and still gain credit. Try to give more detail in your answers.

This answer achieved 4/9

These are not appropriate. They are examples of oil as a fossil fuel after it has been refined.

This is all true, but the essential point is that the rates of formation and use are different.

It is easy to see that the width of the arrows has been measured and the calculation can be followed through to a correct answer.

The first reason uses an energy-density argument and this is a good reason for fossil fuel use. The third reason focuses on ease of construction of engines that use fossil fuels.

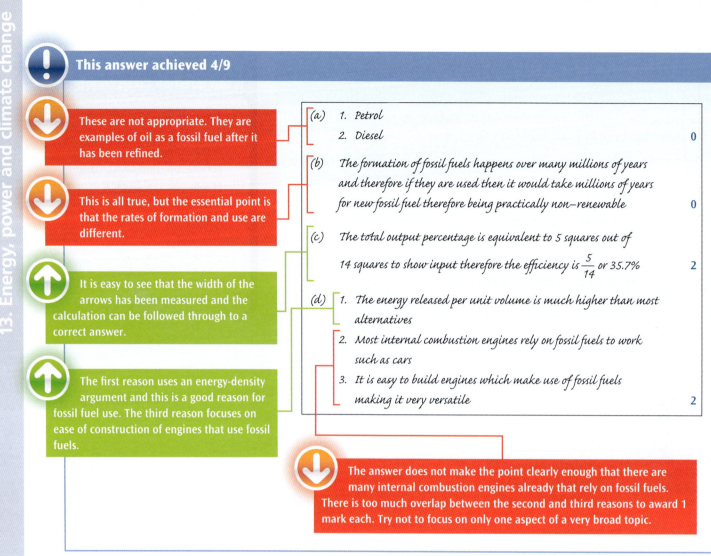

(a) 1. Petrol
2. Diesel   0

(b) The formation of fossil fuels happens over many millions of years and therefore if they are used then it would take millions of years for new fossil fuel therefore being practically non–renewable   0

(c) The total output percentage is equivalent to 5 squares out of 14 squares to show input therefore the efficiency is $\frac{5}{14}$ or 35.7%   2

(d) 1. The energy released per unit volume is much higher than most alternatives
2. Most internal combustion engines rely on fossil fuels to work such as cars
3. It is easy to build engines which make use of fossil fuels making it very versatile   2

The answer does not make the point clearly enough that there are many internal combustion engines already that rely on fossil fuels. There is too much overlap between the second and third reasons to award 1 mark each. Try not to focus on only one aspect of a very broad topic.

This answer achieved 6/9

Although uranium ore is mined, it is not a fossil fuel, as it has not been formed from animal or plant remains.

In principle the fuels could be produced artificially, but it is too expensive to do so.

There is a mark for a good description of the method.

It is not clear where the numbers come from. The total input is related to 14 squares on the grid and it is not clear where the 113 comes from.

Three good and different reasons are given and expressed well.

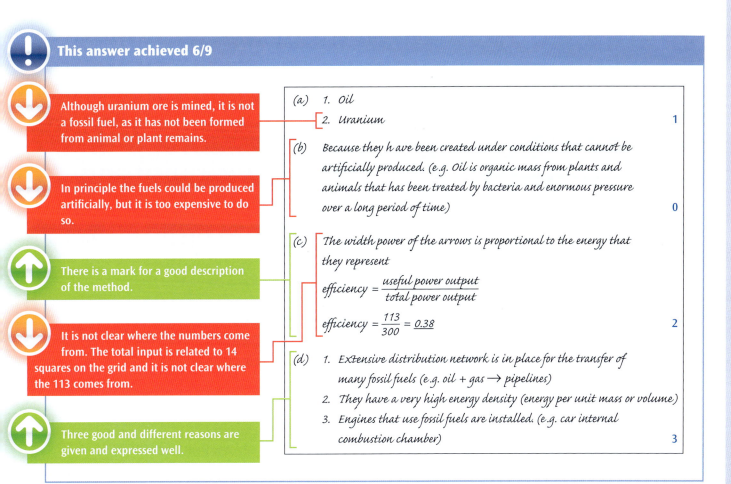

(a) 1. Oil
     2. Uranium   1

(b) Because they h ave been created under conditions that cannot be
     artificially produced. (e.g. Oil is organic mass from plants and
     animals that has been treated by bacteria and enormous pressure
     over a long period of time)   0

(c) The width power of the arrows is proportional to the energy that
     they represent

     $efficiency = \dfrac{useful\ power\ output}{total\ power\ output}$

     $efficiency = \dfrac{113}{300} = \underline{0.38}$ 2

(d) 1. Extensive distribution network is in place for the transfer of
         many fossil fuels (e.g. oil + gas $\rightarrow$ pipelines)
     2. They have a very high energy density (energy per unit mass or volume)
     3. Engines that use fossil fuels are installed. (e.g. car internal
         combustion chamber)   3

14. Option E: Astrophysics

Key terms for this chapter

- luminosity, apparent brightness, apparent magnitude and absolute magnitude
- Wien's (displacement) law and Stefan–Boltzmann law
- Hertzsprung–Russell diagram
- Cepheid variable
- Big Bang model
- critical density, and open, closed and flat universe models
- white dwarf, red giant, red supergiant, supernova, neutron star and black hole
- nucleosynthesis, and Chandrasekhar and Oppenheimer–Volkoff limits
- Hubble's law

Introduction to the universe

You should know:

- the structure of the solar system
- the difference between a stellar cluster and a constellation
- the definition of the light year
- in orders of magnitude, the relative distances between:
  - stars within a galaxy
  - galaxies.

You should be able to:

- explain, in terms of Earth's rotation, the apparent motion of stars and constellations over a period of one night and one year.

Be prepared

- Be able to name the different types of object that make up the solar system: Sun, planets, moons, asteroids, comets.
- It is important to have a sense of the scale of the universe with the relative sizes and separations of objects such as stars, galaxies and stellar clusters.
- Stellar clusters are held together by gravity, whereas a constellation is simply a pattern of stars in the sky imagined by humans on Earth.
- To explain the celestial movement of the stars, imagine you are viewing Earth from a spaceship some way away. The stars are fixed, and Earth is rotating on its axis (in 24 h) and around the Sun (in one year). The night sky viewed by one point on the Earth will be different six months later because the planet will be on the other side of the Sun, and so will now show stars on the opposite side.

Stellar radiation and stellar types

You should know:

- that fusion is the main stellar energy source
- the basic process of fusion in which hydrogen is converted to helium
- the definition of luminosity of a star
- the definition of apparent brightness and how it is measured
- Wien's displacement law
- the system for classifying stars (OBAFGKM).

Stellar radiation and stellar types (continued)

You should be able to:

- explain the equilibrium between radiation and gravitational pressures in a stable stars, including:
- apply the Stefan–Boltzmann and Wien displacement laws
- explain how atomic spectra can be used to deduce chemical and physical data for stars
- describe some different types of stars, including:
  - single star, binary star (spectroscopic and eclipsing)
  - Cepheid
  - red giant, red supergiant, white dwarf
- understand the general regions of a Hertzsprung–Russell diagram.

Be prepared

- Luminosity is the total energy emitted by a star every second.
- Apparent brightness is the total energy per second spread over one square metre at the observation distance. Apparent brightness can be calculated by dividing the luminosity by $4\pi d^2$, where d is the distance from the observer to the star.
- The key to understanding this option is a working knowledge of the Hertzsprung–Russell (HR) diagram.

Do you understand what the axes represent? The axes are logarithmic, and the temperature (star colour) axis is drawn "backwards". You should know how to add the letter classification to the HR diagram, and how to add features to an incomplete diagram.

- Know what is meant by a stable star: the state in which stars spend most of their lives. In this stage, the gravitational pressure making them collapse inwards is balanced by the radiation pressure driving outwards. The star is in a dynamic equilibrium. At the end of its life, one of these processes takes over and produces the imbalance that moves the star to another type.
- Stars act as black-body radiators and have a peak in their intensity–wavelength spectrum. Wien was able to determine the relationship that connects the wavelength of the intensity maximum λ_{max} to the surface temperature T: $\lambda_{max}T = $ constant. Take care with the constant—its value is 2.9×10^{-3} m K. This is metre kelvin, not millikelvin!
- The other important equation is the Stefan–Boltzmann law that relates the luminosity to the star's surface temperature and radius. If the wavelength of the maximum intensity for a star is known, then Wien's law will give the surface temperature, and the temperature and radius will give the luminosity.

Stellar distances

You should know:

- the definition of the parsec
- the definition of absolute magnitude
- that a star's luminosity can be estimated from its spectrum
- that a star's distance can be determined using spectroscopic parallax for distances less than 10 Mpc
- the nature of a Cepheid variable
- the relationship between period and absolute magnitude for a Cepheid variable
- how Cepheid variables can be used as "standard candles".

You should be able to:

- describe and explain the stellar parallax method for determining the distance to a close star
- describe the apparent magnitude scale
- explain how apparent brightness and luminosity can be used to determine stellar distance
- solve problems that involve:
  - stellar parallax

  - apparent brightness, apparent magnitude, absolute magnitude
  - stellar distance
  - luminosity
- use the luminosity–period relationship to determine the distance to a Cepheid variable.

Be prepared

- Several measures of distance are used in astrophysics. Make sure that you can define the parsec (pc) and convert between it and the light year (ly), the astronomical unit (AU) and the kilometre (km). Conversion factors will be given or are in the *Physics data booklet*.
- Apparent magnitude is a logarithmic measure of apparent brightness of stars as seen at the Earth. A factor of 100 gives a change of 5 magnitudes, from 1 (bright) to 6 (dim). The more **negative** a magnitude is, the brighter the star. This magnitude depends on the distance of a star from Earth.
- Absolute magnitude is a measure of the apparent brightness that the star would have if it were 10 pc from Earth. It standardizes the distance for every star, so allows comparisons between magnitudes.

Stellar distances (continued)

- The measurement of stellar distance is complex and depends on the distance of the star from the observer (Earth).

Distance / pc	Method	Brief description of method
0 to 10^2	stellar parallax	Angle subtended across Earth's orbit
0 to 10^4	spectroscopic parallax	Luminosity and apparent brightness are known for two stars (one the Sun). Then if the distance to the Sun is known, the distance to the other can be calculated. Luminosity can be determined from the HR diagram once the star type is known.
0 to 6×10^7	Cepheid variable	The relationship between luminosity and time period of type-II Cepheids is known. Knowledge of period gives luminosity. Measurement of apparent brightness gives distance from inverse-square law.

- Cepheid variables are stars that have a periodic change in luminosity due to the regular contraction and expansion of outer layers of gas at their surface. They are important for the determination of astronomical distance.

Cosmology

You should know:

- the Newton model of the universe
- the Big Bang model of the universe
- the definition of the critical density of the universe
- the differences between open, flat and closed universes
- how the density of the universe determines its development
- the problems associated with determining the density of the universe
- that current scientific evidence indicates that the universe is open.

You should be able to:

- explain Olbers' paradox and show how the Big Bang model resolves the paradox
- describe how space and time originated with the Big Bang
- describe how Penzias and Wilson discovered the cosmic microwave background
- explain how the cosmic background is consistent with the Big Bang model
- discuss, with an example, the international nature of astrophysics research
- understand the issues involved in prioritizing scientific research.

Be prepared

- Newton thought that the universe was static, uniform and infinite in extent, with an infinite number of stars. Olbers realized that this would give a situation in which, wherever one looked in the sky, there would be a star. You should be able to reproduce the simple mathematical argument that Olbers developed.

- Penzias and Wilson discovered the cosmic microwave background radiation (CMB) in the early 1960s and showed that it was consistent with a temperature of 2.8 K for the universe. The CMB is almost homogeneous and arrives from all directions. The modern interpretation of these observations is that the universe originated with a Big Bang—a spontaneous appearance of energy that formed not just material but also space and time. Since that instant, space has been expanding. The consequence is that the wavelengths of the galactic light reaching us have been stretched over the long times during which the light is travelling to Earth. Clearly the universe, under these conditions, is not uniform.

- There are three possible eventualities for the universe: it is open and continues to expand forever; or it is flat and eventually the expansion comes to a halt; or it is closed and will in the future begin to collapse again. These three states are often represented on a diagram

Cosmology (continued)

like the one here. The three lines representing the three possibilities are coincident at the present moment, **not** at the beginning of time. The open, flat or closed outcome depends on the overall density of the universe. A critical value of the density separates the three outcomes: lower than this gives an open state; greater than this means that there is sufficient mass for the gravitational pull to "win" and close the universe again. Present indications are that the universe is open.

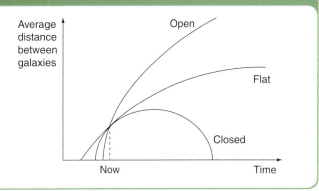

Stellar processes and stellar evolution

You should know:

- the conditions required to initiate fusion in a star
- the effect of the mass of a star on the end product of nuclear fusion
- the changes that occur in the fusion reactions when a star leaves the main sequence and becomes a red giant
- the Chandrasekhar and the Oppenheim–Volkoff limits, and how they predict the fate of stars of different masses
- the characteristics of pulsars.

You should be able to:

- apply the mass–luminosity relation
- compare the fate of red giants and red supergiants
- draw the evolutionary paths of stars on an HR diagram.

Be prepared

- The HR diagram has a major role to play in helping you to learn this section too.
- Fusion, which is the energy conversion process in stars, is the consequence of gravitational attraction acting on the interstellar dust and gas that forms the protostar. As the material condenses, the temperature rises to a point at which electrons are stripped from atoms and a plasma results. The nuclei are now close enough to interact and fuse, with the release of binding energy.
- The protostar's stable state is determined by the mass–luminosity relationship, which is a line with negative

gradient on a standard HR diagram. It will arrive at one point on the main sequence depending on the original amount of gas and dust present in the protostar, and its subsequent history will be determined by this position.

- You need to understand the life histories of stars. One way to learn and revise these is to draw a flow chart beginning with the protostar (interstellar gas and dust) that condenses to form a main-sequence (MS) star. The next steps depend on the mass of the MS star relative to that of the Sun ($M_\odot$):
  - for masses less than $4\,M_\odot$, the star will evolve through a carbon–oxygen core to a planetary nebula and then a white dwarf
  - for masses between $4\,M_\odot$ and $8\,M_\odot$, the same end point is produced via an oxygen–neon core
  - for the largest masses ($>8\,M_\odot$), the end point is a neutron star or black hole reached via an iron core and a supernova.

- The Chandrasekhar limit sets an upper limit of $1.4\,M_\odot$ on the mass of a **white dwarf**. A star larger than this cannot become a white dwarf. The Oppenheimer–Volkoff limit of 2–$3\,M_\odot$ applies to **neutron stars** and gives the limit above which the star can collapse to become a black hole. Try not to confuse these two rules.

- Pulsars are neutron stars that rotate rapidly. They have strong electric and magnetic fields, and can also produce electron–positron pairs as a result of the strength of these fields.

You should know:

- the distribution of galaxies in the universe
- Hubble's law, its limitations and how it is determined
- how Hubble's law can be used to estimate the age of the universe.

You should be able to:

- explain the red-shift of light
- explain how the expansion of the universe made the formation of light nuclei and atoms
- solve problems involving:
  - the red-shift of light
  - recession speed of galaxies
  - Hubble's law.

Be prepared

- Galaxies are grouped into clusters, which are themselves grouped into superclusters.
- Practise applying the equation for the red-shift of light. This equates the fractional shift in wavelength to the fractional speed of the galaxy relative to the speed of light.

- Hubble used this equation, together with the known distances of some galaxies, to establish his law that galactic recession speed $v = Hd$, where d is the galactic distance from the observer and H is the Hubble constant. His problem was lack of accurate knowledge of galactic distance. This is now solved by the use of "standard candles" whose properties are well known. These include Cepheid variables, red and blue supergiants, and globular clusters. If astrophysicists can observe a supernova, they use this too.

- If we assume that the universe expands at a constant rate, then two galaxies presently separated by a distance d will have travelled at a speed Hd to reach that position. This estimates the age of the universe as
  $$\frac{1}{H}$$

- A good revision strategy for the history of the universe is to draw a timeline. For example, at about 10^{-6} s after the Big Bang, neutrons and protons are forming. At 3 minutes, helium nuclei have formed. At 300 000 years, the universe is sufficiently cool for electrons to form atomic hydrogen and helium, and the universe becomes transparent to the photons for the first time. There are other important milestones too.

E1. This question is about the star Antares.

The star Antares is a red supergiant star in the constellation Scorpius.

(a) Describe **three** characteristics of a red supergiant star and state what is meant by a constellation. *[4]*

Red supergiant star:

Constellation:

(b) The apparent magnitude of Antares is $+1.1$ and its absolute magnitude is -5.3.

   (i) Distinguish between apparent magnitude and absolute magnitude. *[2]*

   (ii) Show that the distance of Antares from Earth is 3.9×10^7 AU. *[3]*

   (iii) State the name of the method that is used to measure the distance of Antares from Earth. *[1]*

(c) The apparent brightness of Antares is 4.3×10^{-11} times the apparent brightness of the Sun.

   (i) Define *apparent brightness*. *[1]*

   (ii) Using the answer to (b)(ii), show that Antares is 6.5×10^4 times more luminous than the Sun. *[3]*

[Taken from SL paper 3, time zone 2, May 2009]

How do I approach the question?

(a) At least two characteristics of a red supergiant can be deduced from its name. A third is connected to its colour. Remember that constellations are man-made groupings.

(b) (i) The distinction between the magnitudes is to do with distance from Earth.
(ii) The calculation involves the use of the relationship between apparent magnitude and absolute magnitude. The equation is listed in the *Physics data booklet*.

(iii) Match the distance of Antares with the appropriate method for measuring astronomical distances.

(c) (i) Apparent brightness relates the luminosity to distance between Earth and the star.
(ii) The equation that connects *L* and *d* is in the *Physics data booklet*. Use a ratio approach to establish the answer. Remember that this is a "**show that**" question, and that a high quality of explanation will be required.

This answer achieved 7/14

Three marks are gained for three valid comments about the supergiant.

This answer is just good enough for the mark, but it does not make the point that the stars in a constellation can be very distant from each other.

Magnitudes are not related to luminosity in this way. The absolute magnitude is related to a common distance (10 pc) from the observer.

This answer gets no credit, as it does not attempt to answer the question by a calculation.

Stellar parallax could be used to measure the distance of Antares from Earth.

The definition needs to be given in terms of power rather than luminosity.

There is a correct substitution into the equation for the case of Antares, but the connection to the Sun is not made.

This answer achieved 10/14

A description of the physics of the red supergiant was not asked for in the question. Try not to give unnecessary information, as it wastes your time.

This does not come close enough to the answer to score. The idea that the stars are not distinguishable with the naked eye is odd.

The answer makes the point that apparent magnitude is a comparison as seen from Earth, whereas absolute magnitude is as seen from a distance of 10 pc.

A correct solution is given that takes the calculation to more significant figures than given in the question. This is always a good plan.

This is another possible method for a star at this distance.

This is a correct definition of apparent brightness.

This is a limited answer, in that the correct equation is quoted and used (it has been manipulated) but the substitutions are incorrect.

(a) Red supergiant star:

Heavy enough to fuse element bigger than carbon, low surface temperature due to a very large expansion of the volume (gas law)

Constellation:

Group of stars as seen from Earth which may be very distant from each other in the third dimension, not distinguishable with the naked eye **2**

(b) (i) Magnitude is a number related to brightness, the lower, the brighter the star is. apparent is as seen from Earth and absolute is as seen from a standard distance of 10 parsecs **2**

(ii)
$$\log \frac{d}{10} = \frac{m - \pi}{5} = 1.28 \rightarrow \log d = 1.28 + \log 10$$
$$= 2.28$$

$d = 190.5$ parsecs $= 6 \, U, 2 \, ly = 5.88 \, 10^2 \, m$

$\quad = 3.918 \cdot 10^7 \, AU$

$d = 3.9 \cdot 10^7 \, AU$ **3**

(iii) Spectroscopic parallax **1**

(c) (i) Incident power per unit area (received on Earth) **1**

(ii) $L_A = 64 \, \pi d^2 = 9. \, W^8 \, b_{sum}$

$$L_A = \frac{L_s \cdot 3.9.10^{-11} \times 3.10^{-11}}{4 \, \pi} = 6.5 \cdot 10^4 \, L_s$$ **1**

This answer achieved 12/14

Try not to go beyond the question. It does not ask what the eventual end point of the star will be.

This is a good and complete explanation of what is meant by a constellation.

There is a correct statement that absolute magnitude is related to the star being 10 pc from the observer. However, magnitude is not related to brightness—it was originally a visual scale.

(a) Red supergiant star:

It is a star of very high volume and luminosity that is red in colour. It will very probably collapse into a supernova explosion which will convert it either into a neutron star or a black hole.

Constellation:

It is a collection of stars which are __not__ gravitationally bound. These are grouped like that by human only because they form an interesting shape on the sky. Thus constellation is arbitrary.   **3**

(b) (i) Apparent magnitude is a number associated with the apparent brightness of a star with a small number indicating a high apparent brightness. However, since it not only depends on luminosity but also on distance to the star, the absolute magnitude scale is used, which is the apparent magnitude a star would have when seen from a 10 pc distance from Earth.   **1**

(ii) $m - M = 5 \log\left(\dfrac{d}{10}\right)$

$1.1 - (-5.3) = 5 \log\left(\dfrac{d}{10}\right)$

$6.4 = 5 \log\left(\dfrac{d}{10}\right)$

$\log\left(\dfrac{d}{10}\right) = \dfrac{6.4}{5} = 1.28$

$\dfrac{d}{10} = 19.05$

$d = 19.05 \cdot 10$

$d = 190.5 \ pc$

$190.5 \ pc = 621.18 \ ly$

$621.18 \ ly = 5.88 \cdot 10^{18} m$

$5.88 \cdot 10^{8} m = 3.92 \cdot 10^{7} AU$ **3**

(iii) Spectroscopic parallax since the limit for stellar parallax is 100 pc, and this distance exceeds the limit   **1**

This is a clear and correct definition of apparent brightness.

This is well set out. The cancellations in the substitutions are clearly marked.

(c) (i) Power received (from a star) per unit surface area (on Earth) $b = \dfrac{L}{4\pi d^2} \cdot b = \dfrac{[W]}{[m^2]}$ **1**

(ii) $b = \dfrac{L}{4\pi d^2} \Rightarrow L = b \cdot 4\pi d^2$

$\dfrac{L_A}{L_S} = \dfrac{4.3 \times 10^{-11} \cancel{b_s} \cdot \cancel{4\pi} (3.92 \times 10^7)^2 \cancel{AU^2}}{\cancel{b_s} \cdot \cancel{4\pi} \ AU^2} = 4.3 \cdot 10^{-11} (3.92 \times 10^7)^2$

$= 66075.52$

$= 6.61 \times 10^4 \simeq 6.5 \times 10^4$ **3**

15. Option F: Communications

Key terms for this chapter

- modulation, amplitude and frequency
- bandwidth, sideband
- analogue, digital
- serial, parallel
- time-division multiplexer
- total internal reflection and critical angle
- material and modal dispersion
- operational amplifier
- decibel
- cell and base station

Radio communication

You should know:

- what is meant by modulation
- the difference between a carrier wave and a signal wave
- what is meant by:
  - amplitude modulation (AM)
  - frequency modulation (FM)
- the definition of sideband frequency and bandwidth
- the relative advantages and disadvantages of AM and FM
- block diagram of an AM radio receiver.

You should be able to:

- solve problems involving modulation of a carrier in terms of the amplitude and frequency of an information signal
- understand graphs of the power spectrum of a carrier wave that is amplitude-modulated by a signal with a single frequency.

Be prepared

- You should be able both to describe and to distinguish between amplitude and frequency modulation (AM and FM). In AM, the amplitude of the envelope varies in the same way as the signal wave. In FM, the amplitude of the transmitted signal is constant, and the frequency is modulated. The difference between the instantaneous value of frequency of the carrier wave and the base frequency of the carrier wave is directly proportional to the amplitude of the signal wave.

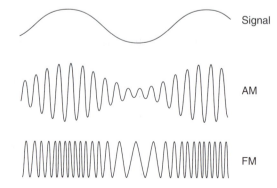

Radio communication (continued)

- Compared with FM, AM suffers from the problem that it is more susceptible to noise. Because electrical discharges change the shape of the AM envelope, these changes will modify the signal. On the other hand, FM receivers expect to receive a constant-amplitude signal and will not reproduce the changes caused to the envelope by noise. FM needs a considerably wider bandwidth than an equivalent AM signal, because the variation in the carrier frequency of the FM is much greater than the variation in the signal frequency.
- Make sure that you can both draw and label all parts in the block diagram of an AM radio receiver. You should be able to state the use of each block.

Digital signals

You should know:

- the difference between analogue and digital signals
- the advantages of digital transmission over analogue transmission
- the principles of the transmission and reception of digital signals, including the functions of modules that are responsible for:
  - sample-and-hold
  - clock timing
  - analogue-to-digital conversion
  - serial-to-parallel conversion
  - parallel-to-serial conversion
  - digital-to-analogue conversion
- what is meant by time-division multiplexing
- the consequences of digital communication and multiplexing on worldwide communication.

You should be able to:

- solve problems involving the conversion between binary and decimal numbers
- explain how the number of bits and the bit-rate affect the reproduction of a transmitted signal
- solve problems involving analogue-to-digital conversion
- discuss moral, ethical, economic and environmental issues arising from access to the internet.

Be prepared

- You should know the steps needed to transmit an analogue signal using digital technology.
- The higher the bit-rate and the greater the number of bits sampled, the better the quality of the transmitted signal.
- You should be ready to carry out calculations on digital conversions, including those involving sampling rates and conversions to and from analogue values.
- Be prepared to write about the social issues that access to the internet brings and also about the consequences of digital worldwide communication.

Optic fibre transmission

You should know:

- what is meant by critical angle and total internal reflection
- how light is transmitted along an optic fibre by means of total internal reflection
- the effects of material dispersion and modal dispersion on the frequency of pulses that can be transmitted
- what is meant by a step-index monomode fibre
- what is meant by attenuation
- what is meant by noise in an optic fibre
- how attenuation in the core of a monomode fibre varies with wavelength
- the role of amplifiers and reshapers in optic fibre transmission.

You should be able to:

- solve problems involving refractive index and critical angle
- solve problems involving attenuation measured in decibels
- solve problems involving optic fibres.

Be prepared

- If light shone into an optic fibre strikes the interior wall at an angle greater than the critical angle, then the light will be totally internally reflected. Given that an optic fibre used for digital transmission has a diameter of tens of micrometres or less, most light will be transmitted down the fibre.
- Be aware of the problems associated with optic fibres, including:
  - material dispersion
  - attenuation
  - modal dispersion.
- Material dispersion is controlled by reshaping the pulse as it travels along the fibre. At separations of about 50 km, the pulses are detected and their edges are squared before retransmission.
- Noise can appear in an optic fibre as a result of stray light entering the fibre at each end. The devices (photodiodes) that convert light energy to an electrical form and vice versa also introduce electrical noise into the system.
- Calculations that may be required include those which involve refractive index and critical angle, and attenuation measured in decibels.

Channels of communication

You should know:

- the possible channels of communication, including:
  - wire pairs
  - coaxial cables
  - optic fibres
  - radio waves
  - satellite communication
- the uses and relative advantages of these possible channels of communication
- what is meant by a geostationary satellite
- the order of magnitude of frequencies used for communication with geostationary satellites
- why the uplink and downlink frequencies used with geostationary satellites are different.

You should be able to:

- discuss the relative advantages and disadvantages for communication of the use of geostationary and polar-orbiting satellites
- discuss the moral, ethical, economic and environmental issues arising from satellite communication.

Be prepared

- For short distances, wire pairs (parallel lengths of wire, sometimes twisted) and coaxial cables (one conductor at the centre of a hollow cylinder made by the other wire) can be used for effective transmission at a wide range of frequencies. Optic fibres can be used for short or long distances. For the greatest distances of transmission, radio waves and satellite communications are used.
- There are two common forms of satellite.
  - Geostationary—these orbit at 42 000 km from the Earth's **centre** above the equator and rotate once every 24 hours.
  - Polar or near-Earth orbiters—these orbit the Earth once every 90 minutes or so, are closer than geostationary satellites and require smaller power for communication.
- Satellites use high transmission frequencies because radio waves up to about 30 MHz are reflected by a layer in the atmosphere and do not reach the satellite. Above this frequency, the frequencies are not reflected by the atmosphere.
- You should be prepared to compare the relative advantages and disadvantages of all these forms of communication.

Electronics

You should know:

- the properties of an ideal operational amplifier (op-amp)
- the use of a Schmitt trigger for the reshaping of digital pulses.

You should be able to:

- draw circuit diagrams for:
  - an inverting amplifier with single input
  - an non-inverting amplifier with single input
  - an op-amp used as a comparator
- derive expressions for the gains of the inverting amplifier and the non-inverting amplifier
- solve problems involving circuits that incorporate operational amplifiers.

Be prepared

- Operational amplifiers are devices that:
  - have two inputs, one that produces an output that is inverted with respect to the input, and one that does not invert the input

  - have a very high input resistance, so that they allow almost no current to be input to the amplifier
  - have a very high open-loop gain—in other words, if there were no modifications made to the amplifier circuit, it would attempt to amplify the input signal to give an output 10^6 or more times greater than the input.
- Although there are many ways to connect an op-amp into an electrical circuit, for your revision you need to concentrate only on these.
  - The single-input inverting amplifier
  - The single-input non-inverting amplifier
  - The two-input comparator
- Ensure that, for all three types of circuit listed above, you can draw or complete a circuit diagram and that you can carry out calculations using the appropriate equations to establish the overall gain of the amplifier circuits.
- The Schmitt trigger (essentially a form of comparator) is used to clean up noisy digital signals and to reshape pulses after they have been affected by dispersion. Use your knowledge of the non-inverting comparator circuit in order to understand how the circuit operates.

The mobile phone system

You should know:

- what is meant by cells and base stations
- that any area is divided into a number of cells, to which are allocated a range of frequencies.

You should be able to:

- describe the roles of the cellular exchange and the public switched telephone network in mobile phone communication
- discuss the use of mobile phones in multimedia communication
- discuss the moral, ethical, economic, environmental and international issues that arise from the use of mobile phones.

Be prepared

- Each cell in the mobile phone network has a base station at its heart. The cell covers an area of a few

square kilometres, and is assigned transmission and reception frequencies that are not the same as those of neighbouring cells. The base station uses these frequencies to communicate with mobile phones within its coverage area, and is connected to a cellular exchange, which allows the phones to communicate with each other. The cellular exchange is also switched into the public switched telephone network (PSTN), so that the mobile phones can connect with wired-in telephones (landlines).

- The base stations detect the movement of a mobile phone from one cell to another, and transfer the phone between each other. The phone is switched via the exchange, so that the connection (if a call is being made) is unbroken.

- You should be ready to discuss the social implications of the use of mobile phones, because, like most new technologies, there are both advantages and drawbacks to their use.

F2. This question is about transmission of signals.

In a particular transmission system a single piece of analogue information is converted into a 4-bit binary "word" represented by the letters ABCD. The word is transmitted along an optic fibre to the receiver. The block diagram shows the principle components for the transmission and reception of this word.

(a) On the diagram label the components X and Y and outline the function of each component. *[3]*

X:

Y:

(b) The diagram is a representation of a two-input time division multiplexer.

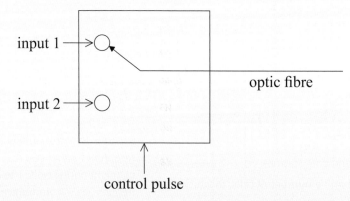

Outline, with reference to the diagram, how this device enables two sets of digital data to be transmitted apparently simultaneously along the same optic fibre. *[2]*

(c) As a signal is transmitted along an optic fibre its signal strength is attenuated. For this reason amplifiers have to be placed at points along the fibre.

(i) Explain what is meant by attenuation. *[2]*

(ii) In a particular fibre, the signal needs to be amplified when the signal power is 8.2×10^{-19} W. The fibre has an attenuation loss of $2.0\,\text{dB}\,\text{km}^{-1}$. Determine, for an input signal of power 5.0 mW, the separation of the amplifiers along the fibre. *[3]*

[Taken from SL paper 3, time zone 2, May 2009]

How do I approach the question?

(a) There are a number of block diagrams mentioned in the syllabus for this option. Make sure that you can identify the meaning of each block, draw them in the correct order, and write a description of the part that each block plays in the overall process.

(b) Time-division multiplexing is an important part of modern telecommunications. You should be aware of why it is used and, as in this case, how the signals are split up to be sent down the fibre.

(c) (i) A statement that the signal gets weaker is not enough—you need to explain in terms of the physics of the signal. You should also indicate why attenuation occurs in a fibre.

(ii) The *Physics data booklet* contains the required equation for this question, but, as usual, you need to be clear about the meaning of the symbols. This is a "**determine**" question, so the working must be clear to the examiner.

This answer achieved 4/10

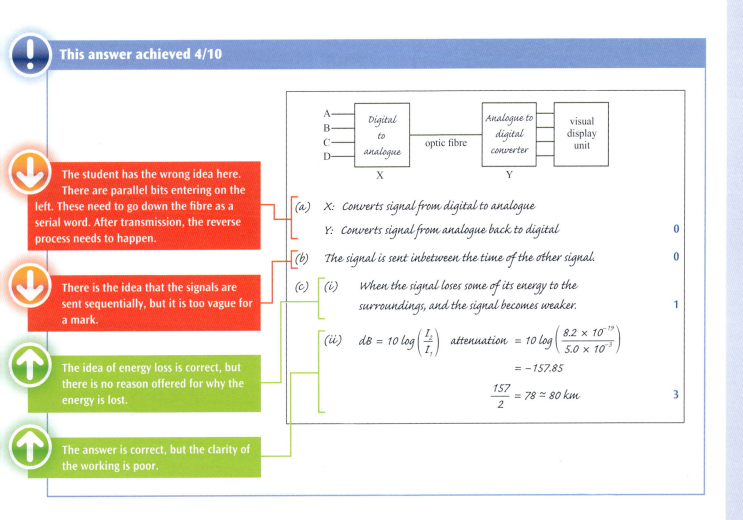

The student has the wrong idea here. There are parallel bits entering on the left. These need to go down the fibre as a serial word. After transmission, the reverse process needs to happen.

There is the idea that the signals are sent sequentially, but it is too vague for a mark.

The idea of energy loss is correct, but there is no reason offered for why the energy is lost.

The answer is correct, but the clarity of the working is poor.

(a) X: Converts signal from digital to analogue

Y: Converts signal from analogue back to digital   0

(b) The signal is sent inbetween the time of the other signal.   0

(c) (i) When the signal loses some of its energy to the surroundings, and the signal becomes weaker.   1

(ii) $dB = 10 \log\left(\dfrac{I_2}{I_1}\right)$ attenuation $= 10 \log\left(\dfrac{8.2 \times 10^{-19}}{5.0 \times 10^{-3}}\right)$

$= -157.85$

$\dfrac{157}{2} = 78 \simeq 80 \ km$ 3

This answer achieved 5/10

The student has the right idea, but the answer is not well expressed and is open to different interpretations. "Puts them down one line with a short time period in between" could mean serial. "One after another" might be better.

There is a reference to the "control pulse" but its exact relationship to the two digital inputs is not clear.

Although there is no explicit reference to energy loss, "dissipation" indicates that the student has an idea of loss of some sort. There is no origin of the loss, however.

There is an error here: 5 mW is 5×10^{-3} W not 5×10^{-6} W.

(a) X: Parallel to serial converter. Takes all the signals and puts them down one line with a short time period in between.

   Y: Serial to parallel converter. Takes all the signals and transmits them at one time. **3**

(b) The pulse switches the input between inputs 1 and 2. There is time between each signal sent by input 1, so during that time input 2 sends a signal. These signals go one after the other through the optic fibre. **1**

(c) (i) It is dissipation along an optical fiber. The signal gets less strong the further it travels. **1**

   (ii) 8.2×10^{-19} W

   2 dB km^{-1} 5 mW input

   $10 \log \dfrac{5 \times 10^{-6}}{8.2 \times 10^{-19}} = \dfrac{247.85 \, dB}{2} = 124 \, km$ **0**

This answer achieved 8/10

This is a good and complete answer that also shows a knowledge of digital signals.

The question says "with reference to the diagram" and there must be a clear description of the role of the control pulse. This is missing.

When a question makes direct reference to a diagram or a previous part of the question, you must follow this lead. The markscheme will be written in these terms, and answers that do not make the necessary reference will be penalized.

The answer indicates what attenuation is and also goes on to show why it happens.

The calculation is clear and well presented. The answer is expressed to an appropriate number of decimal places.

(a) X: A parallel to serial converter changes the information into a string signal of 0s & 1s

   Y: A serial to parallel converter reads the signal and converts the digital string back into A,B,C and D **3**

(b) A part of input 1 is sent down the fibre, and then a part of input 2 is sent down it. The rate of transfer of the optic fibre is large compared to the sampling rate, so it is possible to send 2 signals down the fibre at the same time, and no one will notice as the sampling speed is low anyway **0**

(c) (i) Attenuation is where there is power loss along the fibre due to scattering impurities and absorption of power along the fibre. **2**

   (ii) input power: 5×10^{-3} attenuation: 2 dB km^{-1}

   attenuation $= 10 \log \left(\dfrac{5 \times 10^{-3}}{82 \times 10^{-19}} \right) = 157.85$ dB

   $\dfrac{157.85}{2} \simeq 79 \, km$ **3**

16. Option G: Electromagnetic waves

Key terms for this chapter

- laser—monochromatic and coherent light, population inversion and metastable state
- optical instruments— magnifying glass, compound microscope, astronomical telescope and aberration
- diffraction grating and X-ray crystal diffraction
- two-source interference
- thin-film interference—wedge and parallel films

Nature of EM waves and light sources

You should know:

- the nature of electromagnetic (EM) waves
- what is meant by the dispersion of EM waves in terms of the dependence of refractive index on wavelength
- what is meant by the terms monochromatic and coherent
- that laser light is a source of coherent light
- what is meant by population inversion
- an application of the use of a laser.

You should be able to:

- describe the different regions of the electromagnetic spectrum (the orders of magnitude of frequency and wavelength and a source for each region)
- distinguish between transmission, absorption and scattering of radiation and discuss examples of each
- outline the mechanism for the production of laser light.

Be prepared

- Electromagnetic waves are produced when an electron is accelerated. This results in an oscillating electric field and hence an oscillating magnetic field in a plane at 90° to the electric field. The waves are transverse. In a vacuum, the waves always move at the same speed, whatever their wavelength.

- When moving through a medium, the waves travel at different speeds depending on the wavelength. This gives rise to dispersion, in which waves with different wavelengths travel at different speeds and become spread out in time (and distance). Refraction occurs when a wave enters a new medium.

- Radiation can be transmitted through media or absorbed. Absorption depends strongly on wavelength. As an example, a crystal of sodium chloride (common salt) is transparent to visible light—we can see through it—but is completely opaque to some frequencies of infrared radiation.

- Electromagnetic radiation can also be scattered. Crush a transparent crystal to a fine powder and it will appear white because the light is scattered at different angles as it enters and leaves the individual particles. In the atmosphere, sunlight is scattered by the particles in the air. Blue light is scattered more than red and green, so the sky appears blue, as this colour is distributed more.

Nature of EM waves and light sources (continued)

- Another interpretation of electromagnetic radiation is as a stream of photons. This may help you to understand how the radiation transfers momentum and energy.

- You should be able to state the frequency and wavelength for different regions of the spectrum, together with the way the radiation is usually produced. The regions to concentrate on are: radio, microwaves, infrared, visible, ultraviolet, X-ray and gamma.

- Laser light is monochromatic (contains a very small range of wavelengths) and coherent (all the photons in the light have the same phase).

- The word "laser" stands for "light amplification by stimulated emission of radiation". Ensure that you can explain how a laser works. Key ideas are: how the electrons are initially promoted to a higher level; the importance of the metastable state; population inversion; and the way that one photon can stimulate the production of others.

- Many pieces of modern technology use laser technology, including the CD player and later developments (DVD, BluRay). Make sure that you can describe how these operate.

Optical instruments

You should know:

- for a convex lens, the definitions of:
  - principal axis
  - focal point
  - focal length
  - linear magnification
  - power of the lens
- the difference between a real image and a virtual image
- the definition of the dioptre
- the definitions of:
  - far point for an unaided eye
  - near point for an unaided eye
  - angular magnification for a simple magnifying glass
- for an astronomical telescope in normal adjustment, the equation that relates the angular magnification to the focal lengths of the lenses
- what is meant by spherical and chromatic aberrations when produced by a single lens.

You should be able to:

- construct ray diagrams to locate the image formed by a convex lens
- apply the convention "real is positive, virtual is negative"
- solve problems for a single convex lens using

$$\frac{1}{f} = \frac{1}{u} + \frac{1}{v}$$

- for a simple magnifying glass, derive the expression for the angular magnification for the cases where:
  - the image is formed at the near point
  - the image is formed at infinity

- for a compound microscope:
  - construct a ray diagram with the final image formed close to the near point of the eye
  - solve problems involving the microscope

- for an astronomical telescope:
  - construct a ray diagram with the final image at infinity
  - solve problems involving the telescope:

- describe how spherical aberration in a lens can be reduced

- describe how chromatic aberration in a lens can be reduced.

Be prepared

- Know the basic definitions and, where necessary, be able to sketch a diagram to illustrate your answer.

- Make sure that you really understand the distinction between a real image and a virtual image. A real image can be formed on a screen at the point where rays meet. A virtual image cannot be formed on a screen, and it forms at a point from which diverging rays appear to have come.

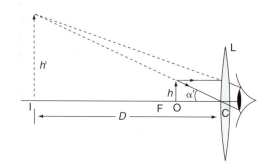

Optical instruments (continued)

- Our eyes have a near point, the closest point on which we can focus, and a far point, the farthest point at which we can see clearly. For a normal eye, the near point is taken to be 25 cm and the far point is at infinity.

- You should be able to draw or complete the ray diagrams for:
  - all the possible object positions for a single convex lens, which will include both cases for the simple magnifying glass (see diagram above)
  - the compound microscope when the final image is formed at the near point of the eye
  - the astronomical telescope when the final image is formed at infinity.

- Practise drawing ray diagrams for all the above cases. Also practise completing the diagram when part of it is already drawn. Some tips for drawing ray diagrams follow.
  - You need at least two "rays" to indicate the position and size of an image. Possible rays start at the tip of an object. One goes parallel to the axis before going through the lens and then through the focal point. The other goes through the centre of the lens and does not deviate at all. Where these two lines meet is the position of the tip of the image.
  - With the telescope and microscope diagrams, remember that it is far easier to fit the diagram on the exam paper if you begin with the lens positions and the final image and draw the ray diagram backwards. Do not mark the focal points in straight away as your choice of image size and lens separation will determine these. The examiner will not know you did it this way! Either way, the key to being able to draw these diagrams is to understand what is happening to the intermediate image inside the instruments.
  - In the telescope, the objective lens forms a real image of the distant object at the focal point of the objective. The telescope is constructed so that this point is also the focal length of the eyepiece lens away from the eyepiece lens itself. So the eyepiece takes this intermediate image and magnifies it again. Draw the intermediate image on your diagram, and then start the construction again to establish the direction of the final image for the whole instrument (the rays will be parallel, as the image is at infinity).
  - For the microscope, the idea is to have the final image near the position of the specimen being magnified. Draw a large final image (it is virtual) and work backwards from this to find where the intermediate image must have been. This image must be closer to

the eyepiece than its focal length. Finally, the objective lens must take the rays from the object and refract them to form this real intermediate image.

How to draw the astronomical telescope diagram in easy steps.

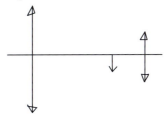

Draw the lenses and the intermediate image.

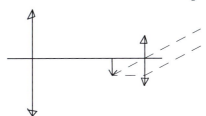

Draw construction lines using the eye lens to show the direction of the final rays.

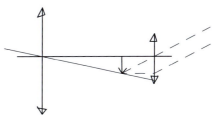

A ray goes through the middle of the objective lens to the top of the intermediate image

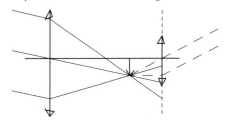

Two other rays are parallel before the lens and then go through the same point.

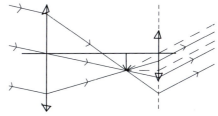

Add the final rays parallel to the construction lines and add the labels—not shown here as an exercise for the student!

Optical instruments (continued)

- For the three optical instruments, it is easier to use the concept of angular magnification rather than linear magnification (because the images are sometimes at infinity):

$$\text{angular magnification} = \frac{\theta}{\theta_0}$$

  where θ is the angle between the principal axis and the ray from the top of the image and θ_0 is the angle between the principal axis and the ray from the top of the object when it is placed at the near point of the eye.

- Calculations in optics usually involve the use of

$$\frac{1}{f} = \frac{1}{u} + \frac{1}{v}$$

  where u and v are the object and image distances, and f is the focal length. Do not forget to take the final reciprocal at the end of the calculation. Linear magnification m is simply

$$m = -\frac{v}{u}.$$

- Be aware of sign conventions. It is important to be systematic about signs and distances in optical calculations. If, in a calculation, a value in a calculation is negative, then this means that the object or image to which the value refers is virtual.

- Aberrations (faults) in lenses come in two forms.
  - Spherical aberrations. In real lenses, rays from the edge are brought to a slightly different focus from those going through the middle. A square object viewed by the lens gives an image with curved sides. The effect can be reduced by restricting the light to the centre of the lens. Alternatively, the lens can be made non-spherical, or two or more lenses can be used.
  - Chromatic aberrations are caused by dispersion. Red and blue wavelengths travel at different speeds in the glass, so are refracted by different amounts and form images at different distances from the lens. This gives rise to coloured fringes at the edges of the lens. The effect is reduced by using two lenses glued together that give opposite aberration effects to each other.

Two-source interference of waves

You should know:
- the conditions necessary to observe interference between two sources
- the intensity distribution for the fringe pattern observed with a double-slit experiment for light.

You should be able to:
- explain the interference pattern produced by waves from two coherent point sources
- outline a double-slit experiment for light
- solve problems involving two-source interference.

Be prepared
- Interference can be observed in all wave types. Sound waves or microwaves can be used to illustrate many of the light phenomena discussed here.
- In order to observe interference between two wave sources, the two waves must be coherent—they must be in phase or have a constant phase difference. This automatically means that they must have the same frequency too. They do not, however, need to have the same amplitude.

- In light experiments, this is done either by using laser light, or by first shining an ordinary light source (a sodium lamp for example) through a single slit. The slit will act as a coherent point source.

- The most basic form of interference is demonstrated in a double-slit experiment. Ensure that you understand both the experiment and the theory. Shining the coherent light through two slits onto a screen some distance away gives a fringe pattern. The pattern has an alternation of dark and light fringes, each fringe parallel to the direction of the slits.

- At a light fringe (a maximum), the rays from the two slits arrive in phase, differing in path length by an integer number of wavelengths—and this phase relationship never varies for this point. The two waves add to give a wave twice the amplitude, four times the intensity of the originals. This is constructive interference. At a dark fringe, the waves arrive half a wavelength (or three half-wavelengths, or five half-wavelengths and so on) out of phase, and always cancel because a crest and a trough arrive together. This is destructive interference.

Two-source interference of waves (continued)

- You may be asked to use the double-slit equation in the exam. It is printed in the *Physics data booklet* as

$$s = \frac{\lambda D}{d}$$

but take care that you know what the symbols mean. You may also be asked to discuss what happens to the slit if the conditions in the experiment change. These changes could include: varying the distance between the slits, varying the distance from slits to screen, varying the wavelength of the light, and varying the width of the slits. You may be asked to sketch the new intensity distribution after the change has been made.

Thin-film interference

You should know:

- how interference fringes are produced by a thin air wedge
- how wedge fringes can be used to measure very small separations
- the condition for light to undergo a π or no (zero) phase change on reflection from an interface
- how a source of light gives rise to an interference pattern when the light is reflected from the top and bottom surfaces of a parallel film
- the conditions for constructive and destructive interference
- how coloured fringes are formed when white light is reflected from thin films
- the difference between fringes formed by a parallel film and a wedge film
- applications of parallel thin films.

You should be able to:

- describe how thin-film interference is used to test optical flats
- solve problems involving wedge films and parallel films.

Be prepared

- The conditions for constructive and destructive interference were discussed in the previous section, and they apply here too.
- When light is reflected between two media and stays in the less dense medium, it undergoes a phase change of π (the wave is inverted). When the light is reflected, staying in the more dense medium, there is no (zero) phase change (0 rad). It is important to remember this in the formation of wedge and parallel-film interference.

- When light travels through a film, there is reflection at both the top and bottom surfaces of the film (see following diagram). When these two reflections subsequently interfere, there will be a path difference caused by the extra distance travelled by light reflected at the bottom surface. If the rays arrive with a phase difference of 180°, they will interfere destructively. An observer looking down on the film in monochromatic light will see an interference pattern. With white light, the pattern will be coloured, because the condition for a minimum is true for only one wavelength corresponding to the colour missing from the white light.

Optical path difference = $2nt$ when $i = 90°$

Angles are shown not equal to 90° incidence for clarity

Refractive index n

- An observer looking down on a film of constant thickness will see a parallel fringe pattern with equal distances between the fringes. For a wedge film, at increasing distances from the start of the wedge, the pattern will get closer together.
- Thin films have a large number of uses. They can be used for non-reflecting surfaces on cameras and other lenses. They can also be used on aircraft to make them less visible to radar. Wedge films can be used to estimate small separations and to test how flat an object is.

Diffraction grating

You should know:

- the effect on the double-slit intensity distribution of increasing the number of slits
- how to use a diffraction grating to measure wavelength.

You should be able to:

- derive the diffraction grating equation for normal incidence
- solve problems involving a diffraction grating.

Be prepared

- A diffraction grating extends the idea of a double slit, but with many more slits. As the number of slits increases, the width of each slit decreases and they become closer together. This spreads the fringe pattern out and concentrates the pattern into a few "orders".

- You should be able to derive and to use the equation appropriate to a diffraction grating. This is

$$2d\sin\theta = n\lambda$$

in the *Physics data booklet*. The idea is that, at very specific points in the pattern, all the rays from all the slits add up to give sharp maxima of intensity. In most other directions, there are so many photons all slightly out of phase with each other that they cancel out.

- The diffraction pattern can be used in many ways, but one common way is for the accurate measurement of wavelength. This relies on the accurate knowledge of the distance between slits in the grating, and this can be established using a microscope with a scale in the eyepiece lens. You should be able to describe how to make this wavelength determination.

X-rays and X-ray diffraction

You should know:

- how X-rays are produced experimentally
- how X-ray diffraction arises when X-rays are scattered in a crystal
- how cubic crystals may be used to measure X-ray wavelengths
- how X-rays may be used to determine the structure of crystals.

You should be able to:

- draw and annotate a typical X-ray spectrum
- explain the features of a characteristic X-ray spectrum
- solve problems involving the accelerating potential difference and minimum wavelength
- derive and solve problems involving the Bragg equation.

Be prepared

- X-rays are produced by making electrons bombard a target made of a heavy metal such as tungsten. The electrons are accelerated by an electric field using a high pd. Most of the electrons lose their energy relatively slowly through glancing collisions with target atoms. The electrons are decelerated and therefore lose their kinetic energy. However, providing the electron energy (and therefore the tube accelerating pd) is high enough, some electrons will strike atoms inelastically and eject electrons from the inner energy levels.

- These two behaviours give rise to the different parts of the X-ray spectrum.

- The continuous spectrum comes from the decelerated electrons. Be able to sketch this spectrum, noting whether you are asked to sketch intensity against wavelength or intensity against X-ray frequency.
- The characteristic spectrum arises from the electrons from higher shells that fall to lower levels, emitting photons of the characteristic frequencies. There are many of these photons, so the characteristic "spikes" on the spectrum are very intense. Add the characteristic spectrum spikes to an intensity graph in approximately the correct positions.

- X-ray diffraction has a number of uses mainly in crystallography: the determination of the molecular structure of new materials. Bragg showed that X-rays can be diffracted by the atomic planes in crystals because the atomic spacing is roughly the same as the wavelength of the X-rays. The angle at which strong diffraction occurs indicates the spacing, and crystallographers can use this information.

- Equally, if a crystal of known spacing is used to diffract X-rays, the wavelength of the X-rays can be measured. The appropriate equation here (the Bragg equation) is

$$2d\sin\theta = n\lambda$$

Make sure that you know the meaning of the symbols and the angle to which this equation refers.

- You should be prepared to carry out X-ray calculations involving the conversion of the X-ray energy to frequency or wavelength, the Bragg equation, and the minimum wavelength that can be produced with a particular accelerating pd.

G2. This question is about interference and lasers.

(a) Two overlapping beams of light from two flashlights (torches) fall on a screen. Explain why no interference pattern is observed. *[3]*

(b) Light from a laser that passes through a double slit is incident on a screen and produces observable interference.
   (i) Outline how the laser produces light. *[2]*
   (ii) State the name of the property that enables the laser light to produce observable interference. *[1]*

(c) Outline how a laser can be used to read the bar-code at the bottom of this page. *[2]*

(d) A plane is flying at $100\,\text{ms}^{-1}$ in a direction parallel to the line joining two identical radio towers as shown in the diagram.

radio towers   200 m

100ms^{-1}

(not to scale)

The two towers each emit a coherent radio signal of wavelength of 5.0 m. The separation of the towers is 200 m. To an observer on the plane the intensity of the received signal goes through a maximum every 5.0 s. Determine the distance from the plane to the line joining the radio towers. *[3]*

[Taken from HL paper 2, time zone 2, May 2009]

How do I approach the question?

(a) This question requires you to use ideas of coherence and to explain why two independent flashlight sources cannot achieve this. In your answer, consider the nature and origin of the light and why it is that this does not achieve coherence.

(b) (i) This is an "**outline**" question, and will not require a detailed answer. With 2 marks, what are the two most important points about the production of laser light?
   (ii) This refers back to part (a).

(c) The two points in this answer will require a reference to what happens to the laser light as it is incident on the bar-code, and how different parts of the bar-code are illuminated.

(d) This is an application of the double-slit interference equation. There are no slits here, but two coherent sources instead. To calculate the fringe spacing, you will need to consider the speed of the plane and the time between maxima.

This answer achieved 4/11

The answer focuses on the differing wavelengths, but the light will be white and contains many wavelengths.

This outline is too confused for full credit. The answer is unclear about the need for a metastable state and for the in-phase release of photons.

The student has not read the question clearly and gives a reasonable answer to a different question, "how is a CD read using laser light".

The answer is correct, but the level of explanation is poor. It is not at all clear how the speed of the plane has been used. However, the interference equation is quoted, but without reference to the meaning of the symbols.

(a)   The reason that there is interference is the beams of light do not have the same wavelength so the path difference is not $n\lambda$ therefore no interference is observed   0

(b) (i)   In order to produce laser light, it's necessary to have to different kinds of molecules, such as He and N. First input electricity so that the molecules are in excited state. Then the He molecules collide with N_2, causing N to be in excited state. When the electron in N drops back, energy is released as the form of light, which are in phase and have the same wavelength.   1

     (ii)   0

(c)   The bar code consists of bumps and flats. The bumps are designed to cause a path difference of the laser light when the light is shined on the code, due to the existence of bumps and flats, there will be flat difference in the laser light and hence it causes interference which can be read by machine   0

(d) $\lambda = 5\,m$ $d = 200\,m$ $x = 5 \times 100 = 500\,m$

$\dfrac{x}{D} = \dfrac{\lambda}{d}$

$\dfrac{5}{200} = \dfrac{500}{d}$

$D = 2 \times 10^4\,m$ 3

This answer achieved 6/11

The point that the two sources are not in phase is present. But the rest of the answer needs too much inference by the examiner.

The statement about population inversion gains a mark, although there is no explanation as to what this means. However, there is no reference to the stimulation of emission in phase.

Coherence is the correct answer for this part.

Although there is a recognition that reflection occurs at the bar-code, there is the suggestion that the bar-code introduces phase differences. This is incorrect.

Full marks are awarded, but this answer is very poorly explained. The 500 m fringe spacing simply appears in the equation.

(a) 1. Because the light from the torches are not monochromatic (they are a mixture of wavelengths).

2. The light emitted from them is not coherent (they give out bursts of light which means the light from the 2 torches will not be in phase.   1

(b) (i) Through a process called population inversion. Normally, electrons are in the lowest energy states in the atom. Population inversion is the act of promoting these electrons to higher energy levels.   1

(ii) Lasers produce monochromatic, coherent light   1

(c) The laser will send coherent, monochromatic waves towards the barcode and some will hit the white sections whereas others will hit the black sections (different heights). Thus, when they reach back (reflected) some will be out of phase with others thus determining the type of barcode.   0

(d) $\dfrac{x}{D} = \dfrac{\lambda}{D}$ $\dfrac{500}{D} = \dfrac{5}{200}$

$D = 20{,}000\ m$

Distance $= 2 \times 10^4\ m$ 3

 The problem with giving little explanation is that an unexplained error can only rarely be given subsequent credit. Had the answer implied that x was 5 km rather than 500 m, then it would not have been clear where this came from, and marks would have been lost.

This answer achieved 9/11

The points about coherence and the random arrival of different wavelengths are correct.

The answer gains full credit with a discussion of population inversion and the release of photons through stimulated emission.

The answer does not make the connection between part (a) and this part.

The ideas that the laser light is reflected where the bar-code is white and absorbed where it is black are correct.

Again, full marks are awarded, but the origin of the 500 m is still not clear even in this able student's work.

(a) *In order for waves to interfere they need to be perfectly in phase; meaning the light is coherent. However, light from a torch is in random bursts of varying wavelengths (colours) and thus the two torches are not coherent with each other and cannot cause intereference.* **2**

(b) *(i)* *Atomic energy levels, when electrons fall from a certain shell they release a certain amount of energy, a "photon". Lasers are set to have all the electrons in a certain shell (by population inversion) and by releasing a photon of coherent energy, causes all the electrons to fall producing coherent light.* **2**

(ii) *Superposition*
But could have been diffraction, population inversion, atomic energy levels etc. **0**

(c) *As white reflects light and black absorbs it, the laser light hitting the white sections will be reflected to the laser, where it can read the missing sections and identify the barcode.* **2**

(d) $\dfrac{x}{D} = \dfrac{n\lambda}{d}$ $\qquad \dfrac{500}{D} - \dfrac{5}{200} = \dfrac{10000}{5} = D$

$$D = 20000 \text{ m}$$ **3**

17. Option H: Relativity

Key terms for this chapter

- frame of reference—inertial and Galilean
- simultaneity
- proper time interval and proper length
- time dilation, twin paradox
- rest mass
- Einstein's principle of equivalence
- black hole and Schwarzschild radius

Introduction to relativity

You should know:
- what is meant by a frame of reference
- what is meant by a Galilean transformation.

You should be able to:
- solve relative velocity problems using the Galilean transformations.

Be prepared
- A frame of reference is a set of axes with clocks at every point at rest relative to an observer.
- The Galilean transformations were those assumed by Newton in his work on mechanics. Frames of reference give the same results in physics experiments if there is uniform velocity of one frame relative to the other. Measurements of acceleration in one frame will give the same results in the other frame too.

Concepts and postulates of special relativity

You should know:
- what is meant by an inertial frame of reference
- the two postulates of the special theory of relativity.

You should be able to:
- discuss the concept of simultaneity.

Be prepared
- An inertial frame of reference is one moving without acceleration (which includes being stationary). Einstein realized that some earlier results for electromagnetism implied that all physical laws must be observed as the same by observers in different frames.
- This leads to the two postulates of special relativity.
  - Physical laws are the same for all inertial observers.
  - All inertial observers observe the same free-space speed of electromagnetic radiation irrespective of observer velocity.
- Remember that an inertial observer is one in an inertial frame of reference.
- An important consequence of the special theory is that events that are simultaneous for one observer are not necessarily simultaneous for a second observer in a different reference frame. You need to practise analysing situations that involve two observers so that you can see clearly and accurately who observes events simultaneously and who does not.

Relativistic kinematics

You should know:

- what is meant by a light clock
- the definition of proper time interval
- the definition of proper length
- how the Lorentz factor varies with relative velocity.

You should be able to:

- derive the time dilation relation
- describe what is meant by length contraction
- solve problems involving time dilation and length contraction.

Be prepared

- A proper time interval is the time interval between events occurring at the same point in space. Note that, although the term "proper time" is commonly used, it always refers to a time interval, not an absolute time. In relativity, time is no longer an absolute. A useful way to show this effect is the graph showing the variation of relative velocity with the factor

$$\gamma = \frac{1}{\sqrt{1 - \frac{v^2}{c^2}}}.$$

Be ready to sketch this graph if necessary.

- Proper length (like proper time interval) is the length of an object when measured by an observer in the same frame of reference as the object (the observer must be stationary with respect to the object).

- A light clock in its simplest form consists of two parallel mirrors with a light beam bouncing between them. The time between photon collisions at the same mirror can be used to time events.

- An observer moving at constant velocity at 90° relative to the clock will see the light undergoing a diagonal track, which will take longer, and the time is therefore dilated.

- Lengths are always contracted in the reference frame of an inertial observer moving relative to the proper length. Here the measured length L is given by

$$L = \frac{L_0}{\gamma}$$

The length is smaller and therefore contracted.

Some consequences of special relativity

You should know:

- the twin paradox and how it is resolved
- the details of the Hafele–Keating experiment and its implications
- the equation that represents the equivalence of mass and energy
- the definition of rest mass.

You should be able to:

- solve problems involving the relativistic addition of one-dimensional velocities
- distinguish between the energy of a body at rest and its total energy when moving
- explain why no object can ever attain the speed of light in a vacuum
- determine the total energy of an accelerated particle.

Be prepared

- In the twin paradox, one of a pair of twins is sent away at high speed to a distant star, turns around, and returns. On return, this twin is now much younger than the Earth-bound twin. The solution is that the situations are not symmetrical, as the twins have not remained in similar inertial frames of reference. One twin has been accelerated, and this has destroyed the symmetry between the twins. Subsequent accelerations of the travelling twin are essential if they are to meet again.

- Hafele and Keating carried out an experiment to test this in 1971 when they sent highly accurate and precise caesium atomic clocks around the Earth in an airliner and on return compared them with identical clocks that had remained in the laboratory. The airliner clocks had "ticked" fewer times, confirming the twin paradox.

- At relativistic speeds, one-dimensional vector velocities have to be added using the relativistic equation. The *Physics data booklet* gives the equation

$$u_x' = \frac{u_x - v}{1 - \frac{u_x v}{c^2}}$$

for this addition. This equation applies when u_x is in the **same** direction as v. If the directions are opposed, then the equation needs two "+" signs.

Some consequences of special relativity (continued)

- The second postulate of the special theory predicts that momentum must be conserved in all inertial frames.

- If an object could attain the speed of light, for an external observer:
  - time would stand still for the object
  - the object would have zero length
  - its energy would become infinite.

- Only light itself can attain the speed of light in a vacuum.

- A common question is to ask about the speed attained by a charged particle being accelerated through a potential difference of V. A way to approach this is to realize that the energy given to the particle can be expressed as

$$eV = mc^2 - m_0c^2$$

or

$$eV = \gamma m_0c^2 - m_0c^2.$$

This allows a solution for γ if everything else is known, and hence the speed can be deduced as a fraction of c.

Evidence to support special relativity

You should know:

- the details and the result of the Michelson–Morley experiment

- how the decay of muons provides experimental evidence to support special relativity

- an experiment that indicates that the speed of light in a vacuum is independent of its source.

You should be able to:

- solve problems involving muon decay.

Be prepared

- Michelson and Morley carried out an interference experiment in 1887 that provided important evidence that there is no absolute reference frame, simply an infinite number of independent inertial frames. You need to know the details of the experiment and how it gave rise to this important null result (a null result is one in which no effect can be detected).

- Muons created tens of kilometres above the Earth are detected at the surface in large numbers. This is surprising because the half-life of the muon is short, and only a few should reach the ground. The value of γ is 5 or more for this situation. In our frame of reference, the time dilation is large and the observed half-life of a muon is about $16\,\mu s$ rather than the expected $3\,\mu s$.

- Gamma rays emitted from neutron sources in distant galaxies are always observed at the same instant even though the photons come from parts of the sources that are moving in different directions relative to Earth. This shows that the speed of the photons is independent of source velocity relative to us.

Relativistic momentum and energy

You should know:

- that momentum can be expressed in the unit $MeV\,c^{-1}$.

You should be able to:

- apply the relation for the relativistic momentum of particles

- apply the formula for the kinetic energy of a particle

- solve problems involving relativistic momentum and energy.

Be prepared

- In non-relativistic mechanics, kinetic energy E_k can be written as $p^2/2m$. The relativistic equivalent of this expression is $E^2 = m_0^2c^4 + p^2c^2$.

- Physicists use what appear to be unusual units for energy, mass and momentum in relativistic calculations. This is because charged particles are often accelerated through potential differences of the order of megavolts (MV). So it is more convenient for them to measure energies in MeV, and this leads to momentum measured in $MeV\,c^{-1}$ and mass measured in $MeV\,c^{-2}$. Make sure you get practice in using these units. You will find the calculations more straightforward without the need to include factors of c^2 and c^4 in your work.

General relativity

You should know:

- the difference between gravitational and inertial mass
- that the principle of equivalence predicts that:
  - light rays will bend in a gravitational field
  - time slows down near a massive body
- the concept of spacetime
- that moving objects follow the shortest path between two points in spacetime
- what is meant by a black hole
- the definition of the Schwarzschild radius
- the concept of gravitational red-shift.

You should be able to:

- describe and discuss Einstein's principle of equivalence
- explain gravitational attraction in terms of the way matter warps spacetime
- calculate the Schwarzschild radius
- solve problems involving:
  - time dilation close to a black hole
  - frequency shifts between different points in a uniform gravitational field.

Be prepared

- Gravitational mass is the mass that is subject to gravitational force, in other words, the m_1 and m_2 terms in
  $$F = \frac{Gm_1m_2}{r^2}.$$
- Inertial mass is the quantity that responds to a force by being accelerated, that is, it is the m in $F = ma$.
- Einstein was able to suggest a principle of equivalence between the above two quantities: gravitational effects cannot be distinguished from inertial effects. According to Einstein, we cannot carry out an experiment that will allow us to decide whether a force arises from a gravitational or an inertial effect, if we cannot see the origin of the force.
- Consequences of the principle of equivalence are that:
  - light is observed to follow a curved path if it is observed by an accelerating observer, but it is also observed to follow a curved path in a gravitational field if the observer is undergoing uniform motion

  - time slows down near a massive body, which becomes clear when you understand that mass warps spacetime.
- The general theory describes gravitational attraction in terms of the warping of spacetime by mass. This is often represented in a three-dimensional way by imagining the universe as a flat rubber sheet. A heavy object (mass) on the sheet deforms the sheet and a small sphere could now be made to "orbit" the object rolling on the curved surface. The path of the object remains the shortest path between two points, but now will take a longer time as the path is no longer the shortest path in the unwarped space.
- A star of large mass and high density can warp spacetime to such an extent that radiation leaving the surface cannot escape. The star has become a black hole. Neither matter nor light can escape from such a star. The largest radius at which this can occur for a particular star is the Schwarzschild radius. You can be expected to calculate this radius for a specified stellar object.
- Time for an object in a region of spacetime near a black hole is dilated relative to an observer. Again, you can be expected to calculate the amount of dilation.
- Gravitational red-shift is another consequence of the principle of equivalence. An observer in a spaceship is close to a source of light, with another, identical, source at the other end of the spaceship. The spaceship and everything in it are accelerating away from the far end. The observer sees the light from the other end of the ship as redder than the light next to him. This is because the light from the far end set out when the spaceship was going slower—and no-one has told the light about the speed change!
- The accelerating spaceship is no different from a stationary spaceship in a gravitational field. Imagine an observer looking down the lift shaft of a building at a lamp. The light of the lamp will be red-shifted by a small amount.
- Many textbooks derive the equation for the frequency change due to gravitational red-shift (it depends on the distance between observer and light source). You do not need to know the derivation of the equation, but you should be able to identify and use it in the exam.

Evidence to support general relativity

You should know:

- an experiment that illustrates the bending of electromagnetic waves by a massive object
- an experiment that provides evidence for gravitational red-shift.

You should be able to:

- describe gravitational lensing.

Be prepared

- You need to know the outline of the following four important experiments for your exam.
- Eddington conducted an experiment during a solar eclipse and was able to confirm the deflection of light by the Sun's gravitational field. This was the first confirmation of Einstein's general theory.

- Shapiro measured a time delay on the round trip of a radar pulse from Earth to Venus and back, and found that it was close to the value predicted by the general theory.
- Pound and Rebka carried out red-shift experiments with sources of gamma rays, one mounted vertically above the other, and found that the lower source was shifted in wavelength by an amount predicted by the general theory.
- Gravitational lensing was first observed in 1979 but has now been seen many times. Light from a distant galaxy is warped by a massive object closer to the observer, resulting in multiple or arc-like images of the distant galaxy.

H3. This question is about gravitational red-shift and black holes.

The concept of gravitational red-shift indicates that clocks run slower as they approach a black hole.

    (a)   Describe what is meant by

        (i)   gravitational red-shift. *[2]*

        (ii)   spacetime. *[1]*

        (iii)   a black hole with reference to the concept of spacetime. *[2]*

    (b)   A particular black hole has a Schwarzschild radius R. A person at a distance of $2R$ from the event horizon of the black hole measures the time between two events to be $10\,\text{s}$. Deduce that for a person a very long way from the black hole the time between the events will be measured as $14\,\text{s}$. *[1]*

[Taken from HL paper 3, time zone 2, May 2009]

How do I approach the question?

(a) (i)   When light moves away from the surface of a planet, it loses energy. What is the implication of this for the frequency of the light?

     (ii) Make sure you are clear what is meant by spacetime. Review the concept of spacetime described above.

(iii) You must describe the black hole in terms of your understanding of spacetime, not any other non-related concept.

(b)   The equation is quoted in the *Physics data booklet*.

This answer achieved 2/6

The description of gravitational distortion is too weak to score marks.

This does not describe spacetime.

A black hole warps spacetime, so this answer is credited with 1 mark.

The student has identified the correct equation and substituted correctly.

(a) (i) Spectral lines are shifted towards the red colour (in its direction) by the gravitational distortion of its spectra — 0

(ii) The plane of the dimension of time and space — 0

(iii) A black hole is very dense and massive such that it warps spacetime — 1

(b) $\dfrac{10\,s}{\sqrt{1-\dfrac{R_s}{2R_s}}} = \dfrac{10\,s}{\sqrt{0.5}} = 14\,s$ — 1

This answer achieved 4/6

The answer is only awarded 1 mark here, as the answer is self-contradictory. The reference to light bending is also wrong in this context. The frequency is in fact lower near a massive body. The reference to Doppler effect is also incorrect.

The student is not awarded a mark here. The mention of a frame of reference is unnecessary.

The idea of warping by the black hole is correct. The idea that the warping is very large is also correct for the second mark.

There is a correct substitution into the correct equation.

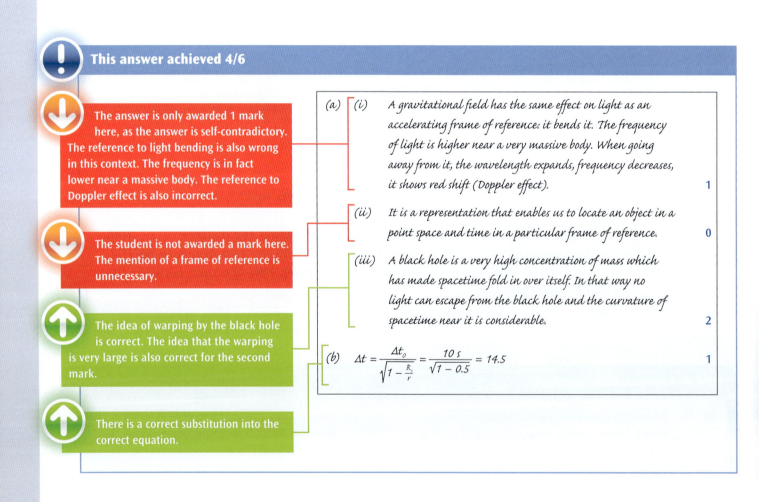

(a) (i) A gravitational field has the same effect on light as an accelerating frame of reference: it bends it. The frequency of light is higher near a very massive body. When going away from it, the wavelength expands, frequency decreases, it shows red shift (Doppler effect). — 1

(ii) It is a representation that enables us to locate an object in a point space and time in a particular frame of reference. — 0

(iii) A black hole is a very high concentration of mass which has made spacetime fold in over itself. In that way no light can escape from the black hole and the curvature of spacetime near it is considerable. — 2

(b) $\Delta t = \dfrac{\Delta t_0}{\sqrt{1-\dfrac{R_s}{r}}} = \dfrac{10\,s}{\sqrt{1-0.5}} = 14.5$ — 1

This answer achieved 6/6

This answer is written in terms of wavelength increase; frequency or wavelength is acceptable.

This is the correct idea, that spacetime is a coordinate system where space and time are combined.

One mark is gained for the idea of mass warping spacetime. One additional mark is awarded for the idea that the warping (curvature) is extreme with a black hole.

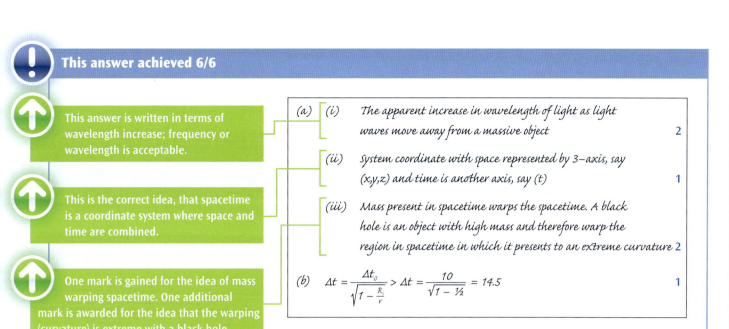

(a) (i) The apparent increase in wavelength of light as light waves move away from a massive object   **2**

(ii) System coordinate with space represented by 3–axis, say (x,y,z) and time is another axis, say (t)   **1**

(iii) Mass present in spacetime warps the spacetime. A black hole is an object with high mass and therefore warp the region in spacetime in which it presents to an extreme curvature **2**

(b) $\Delta t = \dfrac{\Delta t_0}{\sqrt{1 - \frac{R_s}{r}}} > \Delta t = \dfrac{10}{\sqrt{1 - \frac{1}{2}}} = 14.5$ **1**

18. Option I: Medical physics

Key terms for this chapter

- threshold of hearing
- attenuation coefficient and half-value thickness
- acoustic impedance
- nuclear magnetic resonance (NMR)
- exposure, absorbed dose, dose equivalent and quality factor
- half-life—physical, biological and effective

The ear and hearing

You should know:

- the basic structure of the human ear
- how sound pressure variations in the air are changed into larger pressure variations in the inner ear
- the range of audible frequencies experienced by a person with normal hearing
- the definition of intensity and sound intensity level
- the approximate magnitude of the sound intensity level at which people with normal hearing experience discomfort
- why the ear responds logarithmically to changes in intensity
- the effects on hearing of short-term and long-term exposure to noise.

You should be able to:

- solve problems involving intensity levels
- interpret graphs of variation with frequency of intensity level for both normal and defective hearing.

Be prepared

- Learn the accepted frequency range for normal hearing. Sound at a higher frequency than this is called ultrasound.
- Ensure that you have a good working knowledge of the human ear. Be prepared to label a diagram, and to identify individual parts of the ear. You should be able to indicate the regions of the outer, middle and inner ear. Remember to understand the physics of how the ear transfers the sound to the brain, and how, for example, the sound is transmitted into the ear (rather than being reflected back into the air), how the lever system of bones operates, and how this leads to a pressure change.
- The response of the ear is essentially logarithmic. A doubling of perceived loudness corresponds to an intensity increase by a factor of 10.
- Prolonged exposure to high sound intensities leads to hearing loss. This is made clear in graphs of intensity level required for threshold hearing against frequency. Usually, in such graphs, both axes are logarithmic. Be prepared to identify the curves for normal hearing and for hearing that has suffered a loss. The typical symptoms of such loss are a reduction in the sensitivity of the ear and a restriction in the frequency range that a person can detect. This reduction will be over and above the restriction caused by normal ageing.
- Similar graphs are used to plot curves of equal loudness perception. Note that, on all these logarithmic curves, the lower the curve goes, the more sensitive is the ear.

Medical imaging

You should know:

- for X-ray techniques:
  - the definition of attenuation coefficient and half-value thickness, and the derivation of the relation between them
  - recording and display techniques in X-ray detection
  - standard X-ray imaging techniques used in medicine
  - the principles of computed tomography (CT)

- for ultrasound:
  - the principles of the generation and detection of ultrasound using piezoelectric crystals
  - the definition of acoustic impedance
  - the difference between A-scans and B-scans
  - the factors that affect the choice of diagnostic frequency

- for NMR and lasers:
  - the basic principles of NMR imaging
  - examples of the use of lasers in clinical diagnosis and therapy.

You should be able to:

- for X-ray techniques:
  - solve problems using the equation $I = I_0 e^{-\mu x}$

- for ultrasound:
  - solve problems involving acoustic impedance
  - explain why gel is used on the skin

- for NMR and lasers:
  - discuss the use of lasers in pulse oximetry and endoscopy, and as a scalpel and coagulator.

Be prepared

X-ray techniques

- You need to know how medical X-rays are generated using the rotating anode tube.

- There are four ways in which X-rays lose energy as they travel through tissue and bone (attenuation mechanisms).
  - Coherent scattering
  - Compton effect
  - Photoelectric effect
  - Pair production
  You need to understand the underlying physics of these effects.

- The equation that governs the physics of attenuation is $I = I_0 e^{-\mu x}$, where I is the transmitted intensity after the mono-energetic (single-wavelength) X-ray beam has travelled through a thickness x of tissue that has an attenuation coefficient of μ. The incident intensity is I_0. Practise the use of this equation. It has similarities to the radioactive decay equation that you learnt for the core material. In particular, there is the half-value thickness, which is the thickness required to reduce the intensity of the X-ray beam by 50%. This half-value thickness $x_{0.5}$ is given by

$$x_{0.5} = \frac{\ln 2}{\mu}.$$

The similarity between equations in different branches of physics is useful to you in your revision.

- X-rays are not used in medicine "straight out of the tube". After generation, the X-rays are passed through a plate of aluminium to filter out low-energy radiation, and are then collimated (made into a narrow beam) with lead plates. X-rays in the patient are absorbed and scattered. The scattering causes blurring of the image, and this is reduced by using a grid of lead plates to absorb X-rays that are not parallel to the main beam before they strike the imager. The image can be formed on a photographic plate. It is common to use an image-intensifier screen, which contains fluorescent material to improve the brightness of the image and helps to reduce the X-ray dose received by the patient. CCD devices are also used to detect the X-ray image. These are connected to a computer that can analyse the image in real time (see diagram below).

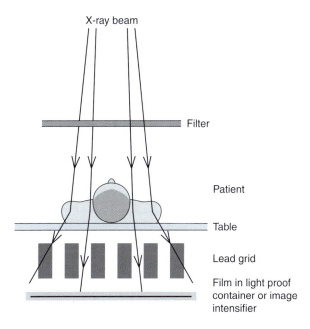

- Computed tomography takes the CCD capture a stage further and enables a 3D view of the patient's organs and

Medical imaging (continued)

- bones to be produced. You need to understand how the technique works, and also what the essential differences are between a CT scan and a 2D X-ray image.

- The quality of an X-ray beam is the term given to its ability to penetrate materials. Quality can be increased by increasing the X-ray tube's accelerating potential difference or beam current, or by increasing the atomic number of the target material.

Ultrasound

- Ultrasound is generated by applying an alternating voltage at a high frequency to a piezoelectric crystal. This crystal then vibrates at the applied frequency. Frequencies used in medical physics vary from less than 1 MHz to 25 MHz or even more.

- The ultrasound technique relies on the differences in the relative amounts of ultrasound reflected from different types of body tissue: bone, fat and so on. Some types reflect well, some badly. The reflected waves reach the detector (usually part of the transmitting device) and a computer produces an image from the resulting return echoes.

- Two types of scan are used, called A-scan (amplitude modulated) and B-scan (brightness modulated).
  - In an A-scan, one transducer (detector) is used and the overall time for the return signal to travel to the body organ and back again is measured. The computer measures these echo pulses and can produce a scan showing a cross-sectional view through the patient.
  - In a B-scan, a linear array of transducers is used simultaneously to scan a plane through the body. After computation, these echo returns can be viewed as a two-dimensional image.

- The choice of diagnostic frequency is a delicate balance for the medical physicist. The greater the frequency, the smaller the wavelength and the better the resolution. However, higher frequencies are attenuated more, and so the returned signals are weaker. Energy is absorbed by tissues, which can cause localized heating. The frequency choice is, therefore, often a compromise.

Nuclear magnetic resonance

- The protons in hydrogen nuclei will align in an external magnetic field so that all the spins are almost entirely lined up. If a patient is exposed to a strong uniform magnetic field, then the hydrogen nuclei in the water molecules in the patient will line up. When the field is removed, the protons revert to a random spin pattern.

- Usually there is one particular region of interest in the patient's body that is of interest to the doctor. An extra, weaker magnetic field—known as the gradient field—is applied to this region. The term "gradient" is used because this weak field has a linear variation with distance through the patient's body.

- There is a characteristic Larmor frequency at which the protons can absorb and then re-emit energy from the magnetic fields by resonance effects. This frequency depends on the strength of the incident magnetic field.

- Because the variation of the strength of the gradient magnetic field with distance is known, a particular Larmor frequency can be associated with a particular position in the body. Information about the strength of the re-emitted radio signal allows the density of protons at a particular point inside the patient to be calculated. This leads to a 3D image of the density of water molecules inside the patient, and hence the shape of various organs.

Lasers

- Laser radiation is monochromatic, collimated and coherent. The source can be pulsed or continuous depending on the application. Various types of laser are used in medicine, with wavelengths ranging from 10 μm to 200 nm and with various intensities.

- In endoscopy, an array of fibres can be used to carry the light into and out of the body. Light can therefore illuminate the interior of the body for internal examinations without the need for surgery. If the relative positions of the fibres taking light out are preserved, then a doctor can view details inside the body from outside. The tube that carries the bundle of fibres can also be used to carry small surgical instruments in order to take biopsies of tissue.

- In surgery, lasers of various intensities and wavelengths are used to carry out surgery of great delicacy. Typical applications include: eye surgery, brain surgery and the removal of skin defects.

- In pulse oximetry, lasers can be used to monitor the oxygen-carrying content of blood. Two lasers, one infrared and one visible light, are shone through the body, and the radiations are partially absorbed by haemoglobin depending on the amount of oxygen in the blood. A computer analyses the results and produces a figure for the percentages of gases present in the blood.

Radiation in medicine

You should know:

- the meanings of radiation dosimetry terms—exposure, absorbed dose, dose equivalent and quality factor (relative biological effectiveness)
- the precautions taken in situations involving different types of radiation
- the concept of balanced risk
- the differences between physical, biological and effective half-lives
- the basis of radiation therapy for cancer.

You should be able to:

- solve problems in radiation dosimetry
- solve problems involving the choice of a radioisotope for a diagnostic or therapeutic application
- solve problems involving particular diagnostic applications.

Be prepared

- There are links between the terms used in dosimetry.
  - Exposure is the charge deposited in tissue per unit mass, and is measured in $C\,kg^{-1}$.
  - Absorbed dose is the energy deposited in tissue per unit mass, and is measured in $J\,kg^{-1}$, which is renamed the gray (Gy).
  - Because not all radiations have the same effect on tissue, it is necessary to have a dimensionless quality factor that represents the effectiveness of a particular radiation.
  - When the quality factor is multiplied by the absorbed dose, the result is the dose equivalent, which is the overall effectiveness of a particular type of radiation on a specific type of tissue.

- Medical personnel who are involved every day with administering radiation to patients need to be protected from harm. This can be done by increasing the distance between the person and the source, introducing a shielding of some kind, and continuously monitoring the radiation of various types received by the medical worker. Ideally, exposure time is kept to a minimum, which is not always possible for medical workers, so film badges and other devices are used to monitor their exposure.

- The concept of radioactive "half-life" is extended when used in a medical context because not only does the radioactive material decay in the usual way, but, as time goes on, the chemical itself is eliminated from the patient's body. The physical half-life T_R is the normal radioactive half-life, which depends only on the nuclear properties of the particular nuclide. The biological half-life T_B gives the time for the isotope to be eliminated from the body by biological processes. These two half-lives can be combined to give the effective half-life T_E, a value that indicates the length of time taken for the activity in the patient's body to halve:

$$\frac{1}{T_E} = \frac{1}{T_R} + \frac{1}{T_B}.$$

- You can be asked to choose an isotope for a particular purpose, given data about a range of nuclides. Remember to consider the requirements to penetrate tissue and to have sensible half-lives for the isotopes. In terms of penetration, tracers that emit alpha or beta particles are unlikely to escape from the body. Half-lives should neither be too short nor too long. However, remember that an isotope with a long half-life has a low activity.

I3. This question is about radio-isotopes.

(a) Distinguish between physical half-life and biological half-life. *[2]*

(b) A person is involved in an accident that causes severe blood loss. Before a blood transfusion can be given it is necessary to measure how much blood has been lost. This can be done by measuring the total volume of blood remaining using an isotope of iodine. Data for two different isotopes of iodine are given in the table.

	Iodine-131	Iodine-123
Physical half-life	8 days	13 hours
Biological half-life	12 days	12 days
Radiation emitted	beta	gamma

Suggest, based on the data given in the table, why iodine-123 is a better choice for use in measuring the total volume of blood in the body. *[3]*

(c) The person in (b) is given an injection of 5.0 ml of a solution containing iodine-123. The initial activity of the solution is 2.5×10^5 Bq. After 0.50 hours it is assumed that the 5.0 ml sample will be distributed evenly throughout the blood system.

The activity of a 5.0 ml sample of the blood after 0.50 hours has an activity of 2.1×10^2 Bq. Deduce that the total blood volume is about 6 litres. *[3]*

[Taken from HL paper 3, time zone 2, May 2009]

How do I approach the question?

(a) The section on "Radiation in medicine" contains some information about the distinction between physical, biological and effective half-lives. Re-read this if necessary.

(b) There are 3 marks to be awarded here and three separate points to make. Consider the activity of the two sources, the relative length of time they will remain in the body, and the ease of detection outside the body.

(c) You need to make an assumption about the extent to which the iodine-123 will have decayed in 0.5 h—you should state this clearly. Its half-life is given in the table in the question. Then the problem becomes a straightforward application of proportion.

This answer achieved 3/8

This is not explained well enough. It implies that the physical half-life is somehow changed in the body.

(a) Physical half-life is the half-life a nucleus would normally have and biological half–life is the half–life of the element when it is inside the human body.

0

The questions contain the command terms **"distinguish"** and **"suggest"**, which are in assessment objectives group 3. This student is giving answers that do not give complete physical statements together with the consequences of these statements.

(b) The physical half–life of iodine-123 is shorter, making the length of time of decay in the body shorter overall. Also, beta particles will be stopped by the tissue and absorbed by the body before reaching the outside.

1

(c) $\dfrac{2.5 \times 10^{5}}{2.1 \times 10^{2}} = 1190.5 \times 5$

$1190.5 \times 5 \times 10^{-3} = 5gs \simeq \underline{6\ litres}$

2

The student does not go on to state the effect on the patient of the shorter time. Neither is the implication of the absorption of beta particles considered.

The calculation is correct and reasonably clear, but the assumption about activity change in 0.5 h is not stated.

This answer achieved 4/8

The statement about biological half-life is too imprecise. It does not emphasize that this process is due to the removal of the chemical from the body.

(a) Physical half-life is the time taken for activity of a substance to halve whereas biological half-life is time taken for the body to halve activity of the substance

1

There is a good idea here in the calculation of the two effective half-lives. There is some credit for this, because on its own, simply saying that iodine-123 is safer to use is not enough. Indeed, as it is more active, it could be described as less safe!

(b) $\dfrac{1}{T_{t}} = \dfrac{1}{T_{B}} + \dfrac{1}{T_{P}}$ For iodine-131, effective half-life is 5 days whereas for iodine-123 it is 12 hours. Therefore it is much safer to use iodine-123 as iodine-131 may cause harm to patient's body

1

(c) $2.5 \times 10^{5} \div 2.1 \times 10^{2} = 1190.5$

$1190.5 \times 5 = 5952\ ml$

$\simeq 6\ Liters$

2

"Deduce" is an assessment objective group 3 command term, and it really is important to state all assumptions and to set work out clearly. Is this a good explanation?

This answer achieved 6/8

This student makes very clear the mechanisms by which the half-life in both cases operates.

The answer considers one aspect and arrives at two conclusions from it. The shorter physical half-life does not technically give less exposure. (Think about what has happened when all the iodine has decayed in both cases, assuming that the same number of atoms were injected into the patient.) But there may be less harm in the light of the second statement.

None of the work here really gets to the heart of the issue of the easier detection of gamma rays outside the body compared to beta particles. So, in principle, a doctor could use less iodine-123 to achieve the same result.

The layout of the answer could be better, as could the level of explanation, but all the required points are here and can be identified by the examiner.

(a) Physical half life is time taken for the amount of radioactive nuclei to reduce by a half by decay. Biological half life is time taken for the amount of radioactive nuclei to reduce by a half by biological processes, such as urination **2**

(b) The physical half-life of Iodine-123 is much shorter than Iodine-131 and so there would be less exposure to the person, i.e. The patient won't be harmed too badly by the radiation

Also, Iodine-123 emits gamma radiation which is far less ionising than beta particles (which is what Iodine-131 emits) **1**

(c) The iodine will not decay very much in 30 mins so I shall assume that the activity does not change to start with.

$$\Rightarrow \frac{2.5 \times 10^5}{2.1 \times 10^2} = 1190.47619$$

$\therefore$ total volume $= 1190.47 \times 0.005$

$$= 5.95 \approx 6L$$ **3**

19. Option J: Particle physics

Key terms for this chapter

- elementary particle
- fundamental interactions—electromagnetic, weak and strong (colour)
- Feynman diagram
- particle accelerators—linear, cyclotron and synchro(cyclo)tron
- particle detectors—bubble chamber, photomultiplier and wire chamber
- quark
- standard model
- lepton, baryon and hadron
- string theory

Particles and interactions

You should know:
- what is meant by an antiparticle
- the Pauli exclusion principle
- what is meant by a Feynman diagram
- know how a Feynman diagram can be used to calculate the probability for a fundamental process
- what is meant by a virtual particle.

You should be able to:
- identify elementary particles and state what is meant by an elementary particle
- describe particles in terms of mass and various quantum numbers such as charge, spin, strangeness, colour, lepton and baryon number
- list the fundamental interactions
- describe the fundamental interactions in terms of exchange particles
- discuss the energy–time uncertainty principle in the context of particle creation
- apply the formula for the range of an interaction where a particle is exchanged
- describe pair annihilation and production through Feynman diagrams
- use Feynman diagrams to predict particle processes.

Be prepared
- An elementary particle is one that has no internal structure. The classes of elementary particles are quarks, leptons and exchange particles (gauge bosons).
  - Quarks come in six flavours and combine to form the heavier particles.
  - Leptons come in six types.
  - Exchange particles come in three groups, and exchange information between quarks and leptons.
- Each particle has an antiparticle. Antiparticles and particles have identical mass and spin, but opposite charge, lepton or baryon number and strangeness (where appropriate). A particle–antiparticle pair can be created given sufficient energy (for example, a gamma photon can annihilate to produce an electron–positron pair in the presence of another mass). Some neutral particles are their own antiparticle.
- Know the link between each fundamental force and its exchange particle(s).
- One way to revise your knowledge of particles and their interactions is to construct a series of tables showing the links between particles and exchange particles.

Particles and interactions (continued)

- Feynman diagrams help to represent the various interactions between quarks, leptons and fundamental particles. In addition, Feynman diagrams are used by particle physicists to calculate the probability of processes occurring. The rules for constructing and interpreting Feynman diagrams are:
  - the diagram is based on a vertex or a series of vertices where fundamental particles come together
  - in this book, time travels from left to right on the diagram (but take care, because other books show time from bottom to top of the page!)
  - an arrow pointing to the right represents a particle moving forwards in time
  - an arrow pointing to the left represents an antiparticle moving forwards in time.

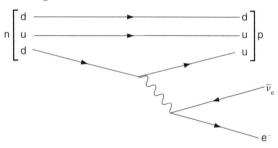

- The Feynman diagram above shows a neutron decaying to a proton and producing a β⁻ (beta minus). The underlying process is that a d quark changes to a u quark. At the first vertex, a d quark enters and a u quark leaves. A W⁻ boson connects to the next vertex, where there is an interaction between the emitted beta particle (e⁻) and an anti-electron neutrino. Notice the arrow direction of the neutrino. The unchanged quarks are added only for completeness.

- Virtual particles exist only during an interaction. They flicker in and out of existence, relying on "borrowing" energy ΔE for their lifetime. This is allowed through the Heisenberg uncertainty principle, and gives an upper limit on the length of time Δt for which the particle can be observed. It also leads to a maximum range R for an interaction. This is because the maximum distance that the particle can move while it exists is $c\Delta t$, which leads to

$$R = \frac{h}{4\pi m_0 c}$$

where m_0 is the rest mass of the virtual particle.

Particle accelerators and detectors

You should know:

- why high energies are required in order to:
  - produce particles of large mass
  - resolve particles of small size.
- the structure and operation, and the advantages and disadvantages, of:
  - a linear accelerator
  - a cyclotron
  - a synchrotron.
- what is meant by bremsstrahlung radiation
- the structure and operation of particle detectors:
  - the bubble chamber
  - the photomuliplier
  - the wire chamber.
- outline the economic and ethical implications and the international aspects of high-energy particle physics.

You should be able to:

- solve problems relating to the production of particles in accelerators.

Be prepared

- The equation $\Delta E = \Delta mc^2$ suggests why a large change in energy ΔE is required to create a particle of large mass Δm.

- To probe a small particle, other particles must be used with a wavelength comparable to the size of the probed object. The governing equation that relates the wavelength of a particle to its energy is $\lambda = h/p$, where λ is the wavelength and p is the momentum. The energy E is equal to $p^2/2m_0$ and therefore a short wavelength is related to a large energy.

- You need to be able to draw, complete or label diagrams of the particle accelerators mentioned in the syllabus. You should also understand how the machines work. They will only accelerate charged particles.
  - Linear accelerators. These are either drift tube accelerators or travelling wave accelerators. The drift tube type is easier to describe and understand.
  - Cyclotrons. These are devices that use two "dees" charged to opposite potentials and switched by a high-frequency alternating pd. As the particles accelerate, their mass increases and they fall out of synchronization with the alternating frequency. This limits the energy that particles

Particle accelerators and detectors (continued)

- can gain and led to the development of the synchro(cyclo)tron. Cyclotrons can be made relatively small and are frequently used for medical purposes.
  - Synchrotron. The most famous are the CERN synchrotron and the TEVAtron at Fermilab. The underground circular tunnel at CERN stretches for 27 km beneath the Jura mountains. Computers are used to control the changing radio-frequency magnetic field to link it to the increasing mass of the circulating particle beam.
- Know the advantages and disadvantages of the three types of accelerator.
  - Linear accelerators are easier to construct than circular machines and do not require large and expensive electromagnets for their operation.

- Cyclotrons have a relatively modest upper energy limit but are cheap to build. They are unsuitable for massive particles.
  - Synchrotrons are very expensive and take large areas of land, but produce the highest energies and the particle beams can be controlled to increase the number of created particles.
- When the particle beams collide with their targets in the accelerator machines, detectors are required to determine the nature of the created particles that stream away from the point of collision. The common types are:
  - bubble chamber
  - photomultiplier
  - wire chamber.

Quarks

You should know:

- the six types of quark
- the quark content of baryons and mesons, including the proton and neutron
- the definition of, and the conservation law of, baryon number
- the colour of quarks and gluons
- the concept of strangeness
- what is meant by quark confinement.

You should be able to:

- deduce the spin structure of baryons and mesons
- explain why colour is necessary in the formation of bound states of quarks
- discuss the interaction between nucleons in terms of the colour force between quarks.

Be prepared

- There are six types of quarks, each with an antiquark. The quarks are: up (u), down (d), strange (s), charm (c), bottom (b) and top (t). The antiquark symbols are written with a bar over the initial. Quarks have a charge of either $-\frac{1}{3}$ of an electronic charge (d, s, b) or $+\frac{2}{3}$ of an electronic charge (u, c, t). All quarks have a baryon number of $\frac{1}{3}$.
- Hadrons are composed of combinations of quarks. As fractional electronic charge is not observed, only certain combinations of quarks are permitted. Mesons are made up of quark–antiquark pairs. Baryons are made from three quarks or three antiquarks.

- The following three conservation laws hold.
  - Charge must be conserved in a particle interaction.
  - Baryon conservation was introduced when some expected reactions were not being observed. Baryons have a baryon number of 1, whereas leptons and mesons have a baryon number of 0. For a reaction to occur, the baryon numbers of the initial particles and the products must be equal.
  - Strangeness conservation is observed in interactions that involve the strong nuclear and the electromagnetic force. It does not need to be conserved in weak interactions.
- Quark colour charge (gluon colour force) arises from the need for the Pauli exclusion principle to be satisfied. Quarks are assigned red, green and blue colours, whereas antiquarks have cyan, magenta and yellow, respectively. Hadrons interact through the strong force and are essentially colourless, as are leptons, which only interact through the weak and electromagnetic forces. Baryons are also colourless, but are composed of coloured quarks. It is believed that, when two baryons are close, the colour force from one baryon can influence the colour charges in the other baryon. Coloured particles exchange colour combined gluons during the strong interactions, and therefore the quark colours can change.
- Free quarks have never been observed, and quark confinement is therefore thought to occur. The suggestion is that the energy increases with increasing separation (in contrast to electromagnetic forces and energies, which decrease as r increases). No four-quark entities have been discovered, and so only colour-neutral particles appear to exist.

Leptons and the standard model

You should know:

- the three-family structure of quarks and leptons
- the lepton number of the leptons in each family
- the significance of the Higgs particle (Higgs boson)
- that strangeness is conserved in all strong and electromagnetic interactions, but not always in weak interactions.

You should be able to:

- solve problems in particle reactions by using conservation laws for:
  - charge
  - total energy
  - momentum
  - baryon and family lepton numbers.

Be prepared

- The standard model suggests that there are three families of objects from which all interactions and particles can be constructed: quarks, leptons and neutrinos. Each particle in these families has an equivalent antiparticle.
- Higgs suggested that the mass of a particle results from its electroweak interactions with a Higgs field. This theory is likely to be confirmed if a Higgs boson can be discovered. The search for this high-mass boson is currently running at CERN, and it may be that this particle will be identified over the next few years.

Experimental evidence for the quark and standard models

You should know:

- what is meant by:
  - deep inelastic scattering
  - asymptotic freedom
  - neutral current
- how the existence of a neutral current is evidence for the standard model.

You should be able to:

- analyse the results of deep inelastic scattering experiments.

Be prepared

- Deep inelastic scattering occurs when highly energetic particles are made to collide with nucleons. The particles, acting as very short-wavelength waves, are able to be diffracted by the quarks. Because the quarks cannot exist alone, more quarks are brought into existence and a stream of new particles is created.
- The unusual property that, as quarks separate, the energy increases was mentioned earlier. As the quarks approach each other, the energy stored drops asymptotically to zero. This property is known as asymptotic freedom.
- Originally the W^+ and W^- bosons were thought to be responsible for radioactive decay. Later, an electroweak theory suggested a unification of the electromagnetic and weak forces. If this was to be the case, there would need to be a neutral charge carrier for interactions in which no charge transfer occurs. Such an interaction is known as a neutral current reaction. It was eventually discovered in 1983 and the particle involved was named the Z^0. The W^+ and W^- always involve transfer of charge.

Cosmology and strings

You should know:

- the order of magnitude of the temperature change in the universe since the Big Bang
- that the early universe contained almost equal numbers of particles and antiparticles
- the (non-quantitative) theory of strings.

You should be able to:

- solve problems involving particle interactions in the early universe
- suggest a mechanism by which matter has come to dominate over antimatter.

Be prepared

- You may find it helpful to draw a graph of the temperature of the universe against time and to annotate this with details of the processes occurring at that time. Use a logarithmic scale on both axes. The following table will provide a starting point for this graph. Use your books, notes and the internet to establish the details of the processes occurring in each era.

Cosmology and strings (continued)

Era begins at about	Era name	Approximate initial temperature / K
10^{-43} s	grand unified	10^{32}
10^{-35} s	hadron	10^{28}
10^{-4} s	lepton	10^{15}
1 s	radiation	10^{10}
3 minutes	nucleosynthesis	10^{9}
10^{6} years	matter	10^{4}
10^{10} years	present	3

the photons was so great that both pair production of particles and antiparticles and their annihilation were common when the age of the universe was about one second. However, soon after, annihilation became the predominant mechanism, and the numbers of particles and antiparticles fell dramatically, leaving a slight imbalance of particles. The ratio froze at its present value when the photon energy fell below the pair-production threshold.

- The theory of gravity is not at present reconciled with quantum theory. One present way forward is the continuing development of string theory, in which the concept of a fundamental particle is replaced by that of an oscillating string of length about 10^{-35} m existing in other dimensions beyond the three that we sense. The modes of vibration of the strings determine the properties of the "particles" that they represent.

- The early universe contained almost equal numbers of particles and antiparticles. This is because the energy of

J1. This question is about fundamental interactions and elementary particles.

(a) In the table identify the exchange particle(s) associated with the two fundamental interactions given. *[2]*

Interaction	Exchange particle(s)
Electro-weak	
Strong	

(b) State why the exchange particles are known as elementary particles. *[1]*

(c) An exchange particle associated with the weak interaction has a mass of about $90\,\text{GeVc}^{-2}$.

Estimate the life-time of the particle. *[3]*

(d) The diagram is a Feynman diagram that represents the strong interaction between quarks.

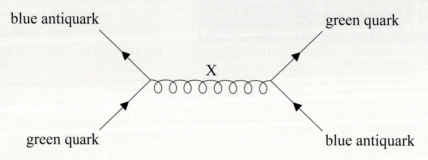

(i) Identify the exchange particle X. [1]

(ii) Explain why the quarks have a colour associated with them. [2]

[Taken from HL paper 3, time zone 2, May 2009]

How do I approach the question?

(a) You should learn the exchange particles for all the interactions.

(b) Explain what is meant by a fundamental particle and why an exchange particle fits this description.

(c) The appropriate equation in the Physics data booklet indicates the relationship between ΔE and Δt. You will need to use Einstein's mass equation to change E to m. Take care with the units.

(d) (i) There is an exchange particle involved with the strong interaction between quarks.

(ii) If there were to be no colour force, there would be no way to distinguish some quarks. What problem would arise in this case?

This answer achieved 3/9

Baryons are not exchange particles.

The examiner ignores the (incorrect) reference to pi-mesons.

There may be a word missing here. But elementary particles are not made of anything smaller.

The idea is good, but the units are not treated correctly.

A less vague answer is required.

The reference to the Pauli exclusion principle scores 1 mark.

(a)

Interaction	Exchange particle(s)	
Electro–weak	*leptons Baryons*	
Strong	*π–mesons, gluons*	1

(b) *They are known to have anything contained inside of them* — 0

(c) *based on* $\Delta E \Delta t \geq \frac{h}{4\pi}$ $\therefore \Delta\ 90\ GeVc^{-2}\ \Delta t \geq \frac{6.63 \times 10^{-34} kg}{4\pi}$

$= \frac{90}{90} GeVc^{-2}\ \Delta t \geq \frac{5.275 \times 10^{-35}}{90}$

$\therefore t \geq 5.86 \times 10^{-37}$ *seconds* — 1

(d) (i) *Virtual particle* — 0

(ii) *The colour is used to classify the various quarks. The six quarks are paired with a counterpart as stated by Pauli exclusion principle. The combination of colour ie red blue green or red and antigreen have to = white ∴ the combination is possible* — 1

This answer achieved 6/9

The Z^0 is not the only particle in the electro-weak interaction. A more complete list is required.

Gluon is correct for the strong interaction.

The student understands the need to convert energy to mass, but the conversion is incorrect and omits the conversion from GeV.

The reference to Pauli exclusion gains 1 mark.

(a)

Interaction	Exchange particle(s)	
Electro–weak	Z^0	
Strong	*gluon*	1

(b) *Elementary particles do not have internal structure. Exchange particles do not have internal structure therefore they are elementary particles.* — 1

(c) $\Delta E\ \Delta t \geq \frac{h}{4\pi}$ $\qquad E = mc^2$

$\qquad\qquad\qquad\qquad E = 90 \times 10^9 \times (3 \times 10^8)^2$

$8.1 \times 10^{27}\ \Delta t \geq \frac{h}{4\pi}$ $\qquad E = 8.1 \times 10^{27}$

$8.1 \times 10^{27}\ \Delta t \geq 5.28 \times 10^{-35}$

$\Delta t \geq 6.52 \times 10^{-63}\ s$ — 2

(d) (i) *Gluon* — 1

(ii) *The colour of quarks is a quantum number introduced to differentiate them in order to satisfy Pauli's exclusion principle* — 1

This answer achieved 9/9

The responses are appropriate. The omission of the signs for W and Z is ignored and allowed as benefit of the doubt.

The estimate is clear and the conversions to mass are accurate. The answer is correct (although, as this is an estimate, fewer significant figures in the final answer would have been better).

The extra information here is sufficient to attract 2 marks.

(a)

Interaction	Exchange particle(s)
Electro–weak	photon, W, Z
Strong	gluon

2

(b) Because they are not made from internal constituents or believed to have an internal structure.

1

(c) $\Delta E \, \Delta t \geq \dfrac{h}{4\pi}$ $h = 6.63 \times 10^{-34}$

$90 \times 10^4 \text{ eVc}^{-2}$ $\dfrac{h}{4\pi} = 5.276 \times 10^{-35}$

$(90 \times 10^9) \times (1.6 \times 10^{-19}) = 1.44 \times 10^{-8}$

$\therefore \Delta t = \dfrac{5.276 \times 10^{-35}}{1.44 \times 10^{-8}}$

$\Delta t \geq 3.66 \times 10^{-27} \text{ s}$

3

(d) (i) Gluon

1

(ii) Colour is necessary to satisfy the Pauli exclusion principle. Gluons have a colour – anti-colour nature in order to exchange colour between different quarks. Hadrons have no colour because they are confined

2

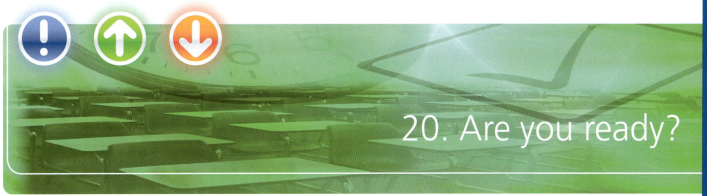

20. Are you ready?

I hope that this book has helped you and that you will continue to use it to prepare for your physics exams.

Here is a set of past exam papers from May 2010 for you to practise your skills. It is your choice whether to answer these against the clock under exam conditions or to answer one question at a time.

Paper 1

4. Which of the following is a correct statement of Newton's second law of motion?

   A.   A force acting on a body is proportional to the mass of the body.

   B.   The rate of change of momentum of a body is equal to the net external force acting on the body.

   C.   The momentum of a body is proportional to the net external force acting on the body.

   D.   A force acting on a body is proportional to the acceleration of the body.

10. Water at a temperature of $0\,^{\circ}\mathrm{C}$ is kept in a thermally insulated container. A lump of ice, also at $0\,^{\circ}\mathrm{C}$, is placed in the water and completely submerged.

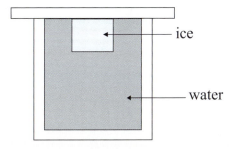

   Which of the following is true in respect of both the net amount of ice that will melt and the change in temperature of the water?

	Net amount of ice that melts	Change in temperature of water
A.	all will melt	no change
B.	some will melt	decrease
C.	none will melt	no change
D.	all will melt	decrease

12.   A gas is contained in a cylinder by a piston.

The gas is compressed rapidly by moving the piston in the direction shown. The best explanation for the resulting increase in temperature of the gas is that the molecules of the gas gain kinetic energy

A.   from the moving piston.

B.   by colliding more frequently with each other.

C.   by being pushed closer together.

D.   by colliding more frequently with the walls of the cylinder.

15.   The graph shows how the velocity v of an object undergoing simple harmonic motion varies with time t for one complete period of oscillation.

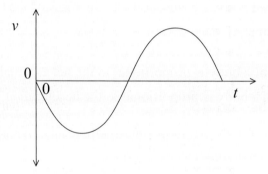

Which of the following sketch graphs best shows how the total energy E of the object varies with t?

A.

B.

C.

D.

18. The fundamental (first harmonic) frequency of the note emitted by an organ pipe closed at one end is f. What is the fundamental frequency of the note emitted by an organ pipe of the same length that is open at both ends?

A. $\dfrac{f}{4}$

B. $\dfrac{f}{2}$

C. $2f$

D. $4f$

21. The images of two sources are just resolved. Which of the following is a correct statement of the Rayleigh criterion for this situation?

A. The central maximum of the diffraction pattern of one source must coincide with the central maximum of the diffraction pattern of the other source.

B. Light from the sources must pass through a circular aperture.

C. Light from the sources must be coherent.

D. The first minimum of the diffraction pattern of one source must coincide with the central maximum of the diffraction pattern of the other source.

25. A copper sheet is suspended in a region of uniform magnetic field by an insulating wire connected to a horizontal support. The sheet is pulled to one side so that it is outside the region of the field, and then released.

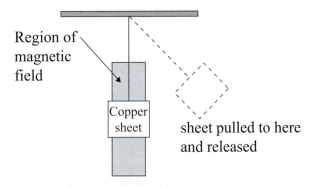

The uniform magnetic field is directed into the plane of the paper.

Which of the following is true for both the direction of the induced current in the sheet and the change in amplitude of the oscillations of the sheet with time?

	Direction of induced current	Change in amplitude
A.	stays the same	no change
B.	changes	decreases
C.	stays the same	decreases
D.	changes	no change

30. Which of the following is an assumption of the Schrödinger model of the hydrogen atom?

A. The orbital path of the electron fits a standing wave.

B. The position of the electron is undefined but its momentum is well defined.

C. The momentum of the electron is undefined but its position is well defined.

D. The electron is described by wavefunctions.

33. For the five-bit binary number 11010, which of the following correctly identifies both the most-significant bit (MSB) and the equivalent decimal number?

	MSB	Decimal
A.	1	52
B.	0	26
C.	1	26
D.	0	52

36. Degraded energy is energy that is

A. stored in the Earth's atmosphere.

B. available from non-renewable energy sources.

C. converted into work in a cyclical process.

D. no longer available for the performance of useful work.

39. Which of the following correctly shows the energy change in a photovoltaic cell and in a solar heating panel?

	Photovoltaic cell	Solar heating panel
A.	solar → electrical	solar → thermal
B.	electrical → thermal	solar → electrical
C.	solar → electrical	electrical → thermal
D.	electrical → thermal	solar → thermal

Paper 2

Section A

A1. This question is about electrical resistance.

The graph shows the variation with temperature T of the resistance R of an electrical component.

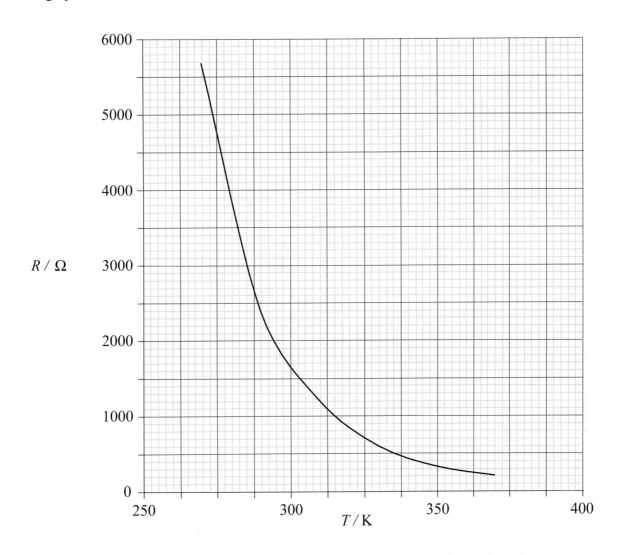

(a) A student hypothesizes that the resistance is inversely proportional to the temperature. Use data from the graph to show whether the hypothesis is supported. *[3]*

(b)   A second student suggests that the relationship is of the form

$$\lg R = a + \frac{b}{T}$$

where a and b are constants.

The student plots the graph below. Error bars have been included for the sake of clarity.

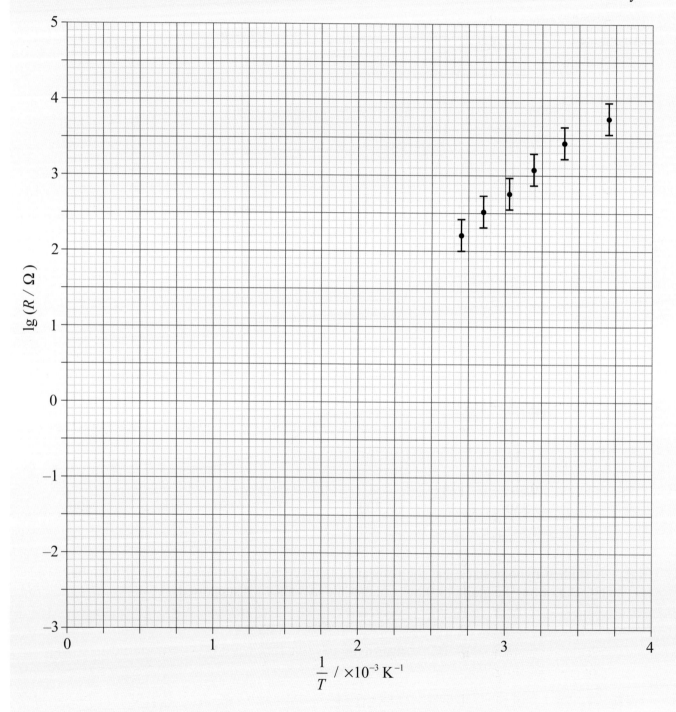

(i)   Explain how the graph drawn could be used as evidence to support the student's suggestion.

[2]

(ii)   Use the graph to determine the constants *a* and *b*.   [4]

   a:

   b:

(iii) Using your answers to (b)(ii), determine a value for the resistance of the component
       at a temperature of 260 K.   [2]

A2.   This question is about forces.

An athlete trains by dragging a heavy load across a rough horizontal surface.

The athlete exerts a force of magnitude *F* on the load at an angle of 25° to the horizontal.

(a)   Once the load is moving at a steady speed, the average horizontal frictional force acting
      on the load is 470 N.

      Calculate the average value of *F* that will enable the load to move at constant speed.   [2]

(b)   The load is moved a horizontal distance of 2.5 km in 1.2 hours.

      Calculate

      (i)   the work done on the load by the force *F*.   [2]

      (ii)   the minimum average power required to move the load.   [2]

(c)   The athlete pulls the load uphill at the same speed as in part (a).

      Explain, in terms of energy changes, why the minimum average power required is greater
      than in (b)(ii).   [2]

A3.   This question is about solar heating panels.

(a)   A village consists of 120 houses. It is proposed that solar panels be used to provide hot
      water to the houses.

      The following data are available.

      average power needed per house to heat water   = 3.0 kW
      average surface solar intensity   = 650 W m^{-2}
      efficiency of energy conversion of a solar panel = 18%

      Calculate the minimum surface area of the solar panels required to provide the total power
      for water heating.   [3]

(b)   Suggest **two** disadvantages of using solar power to provide energy for heating water.   [2]

      1:

      2:

A4. This question is about a generator.

(a) Define *electromotive force*. *[1]*

(b) The graph shows the variation with time of electromotive force (emf) for a generator.

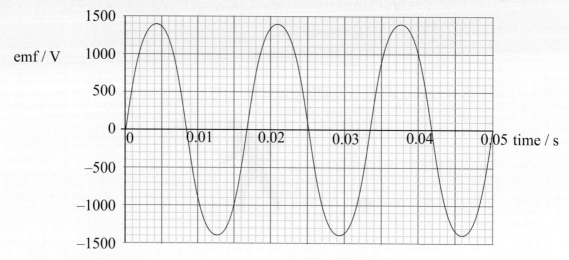

(i) Calculate the rms value of the emf of the generator. *[2]*

(ii) The speed of rotation of the generator is halved with no other changes being made. On the graph, sketch the variation of emf with time. *[2]*

(iii) Explain why the graph you drew in (ii) is different from the original graph. *[2]*

A5. This question is about an ideal gas.

(a) The graph shows a pressure-volume (*P–V*) relationship for a fixed mass of an ideal gas.

The gas undergoes a three-stage cycle AB, BC and CA.

(i) Identify the isochoric (isovolumetric) change of state. *[1]*

(ii) Use data from the graph to show that the change AB is isothermal. *[3]*

(iii) State the **two** changes for which thermal energy is transferred from the gas. *[1]*

(iv) Calculate the work done by the gas in the change BC. *[2]*

(b)   The amount of gas in part (a) is 0.74 mol.
      Calculate the maximum temperature of the gas during the cycle in part (a).   [4]

(c)   The fixed mass of gas is heated from temperature T_1 to T_2 at constant volume. Explain why, if this fixed mass of gas is heated from T_1 to T_2 at constant pressure, the amount of energy required is different.   [3]

Section B

*This section consists of four questions: B1, B2, B3 and B4. Answer **two** questions.*

B1.   This question is in **two** parts. **Part 1** is about solar radiation. **Part 2** is about charge-coupled devices (CCDs).

Part 1   Solar radiation

(a)   State the Stefan–Boltzmann law for a black body.   [2]

(b)   The following data relate to the Earth and the Sun.

Earth-Sun distance	$=1.5\times10^{11}\,m$
radius of Earth	$=6.4\times10^{6}\,m$
radius of Sun	$=7.0\times10^{8}\,m$
surface temperature of Sun	$=5800\,K$

      (i)   Use data from the table to show that the power radiated by the Sun is about $4\times10^{26}\,W$.   [1]
      (ii)   Calculate the solar power incident per unit area at a distance from the Sun equal to the Earth's distance from the Sun.   [2]
      (iii) The average power absorbed per unit area at the Earth's surface is $240\,W\,m^{-2}$. State **two** reasons why the value calculated in (b)(ii) differs from this value.   [2]
      (iv)   Show that the value for power absorbed per unit area of $240\,W\,m^{-2}$ is consistent with an average equilibrium temperature for Earth of about $255\,K$.   [2]

(c)   Explain, by reference to the greenhouse effect, why the average temperature of the surface of the Earth is greater than $255\,K$.   [3]

(d)   Suggest why the burning of fossil fuels may lead to an increase in the temperature of the surface of the Earth.   [3]

Part 2   Charge-coupled devices (CCDs)

(a)   List **two** advantages of using CCDs as compared to using film for image capture.   [2]
      1:
      2:

(b)   A CCD used to capture X-ray images consists of a square $26\,000\times26\,000$ arrangement of pixels. Each pixel produces 16 levels of information. Images captured by the CCD are stored on a disk which can store a maximum of 4.0×10^{11} bits of data.
      Determine the number of images that may be stored on one disk.   [3]

(c)   Outline
      (i)   how X-rays incident on a pixel in a CCD produce a voltage signal.   [2]
      (ii)   how voltage signals such as that in (c)(i) are used to store the X-ray information.   [3]

B2. This question is in **two** parts. **Part 1** is about water waves. **Part 2** is about projectile motion.

Part 1   Water waves

A small sphere, mounted at the end of a vertical rod, dips below the surface of shallow water in a tray. The sphere is driven vertically up and down by a motor attached to the rod. The oscillations of the sphere produce travelling waves on the surface of the water.

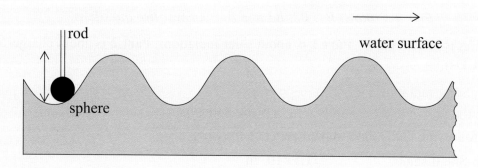

(a)   State what is meant by a travelling (progressive) wave.   *[1]*

(b)   The diagram shows how the displacement of the water surface at a particular instant in time varies with distance from the sphere. The period of oscillation of the sphere is 0.027 s.

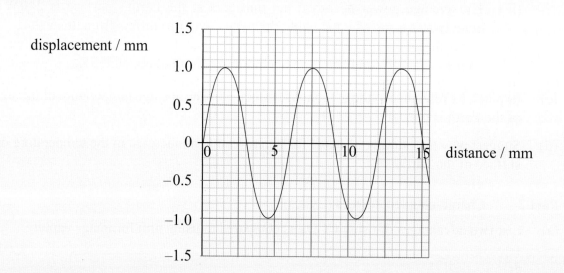

Use the diagram to calculate, for the wave,

(i)   the amplitude.   *[1]*

(ii)   the wavelength.   *[1]*

(iii)   the frequency.   *[1]*

(iv)   the speed.   *[1]*

(c)   The wave moves from region A into a region B of shallower water. The waves move more slowly in region B. The diagram (not to scale) shows some of the wavefronts in region A.

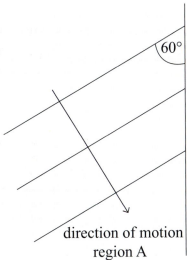

60°

direction of motion
region A   region B

(ii) Theory suggests that the wave speed c is related to the water depth d by

$$c = \sqrt{gd}$$

where g is a constant.

The refractive index for waves travelling from region A to region B is 1.4.

Determine the following ratio.

$$\frac{\text{water depth in A}}{\text{water depth in B}}$$ *[3]*

(d)   Another sphere is dipped into the water. The spheres oscillate in phase. The diagram shows some lines in region A along which the disturbance of the water surface is a minimum.

lines of minimum
disturbance

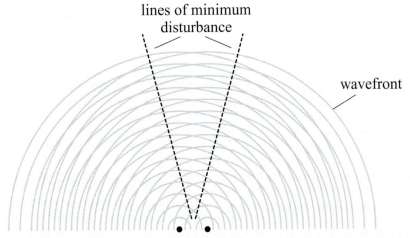

wavefront

(i)   Outline how the regions of minimum disturbance occur on the surface.   *[3]*

(ii) The frequency of oscillation of the spheres is increased.

State **and** explain how this will affect the positions of minimum disturbance.   *[2]*

Part 2   Projectile motion

A sphere is projected horizontally. The sphere is photographed at intervals of 0.10 s. The images of the sphere are shown against a grid on the diagram. Air resistance is negligible.

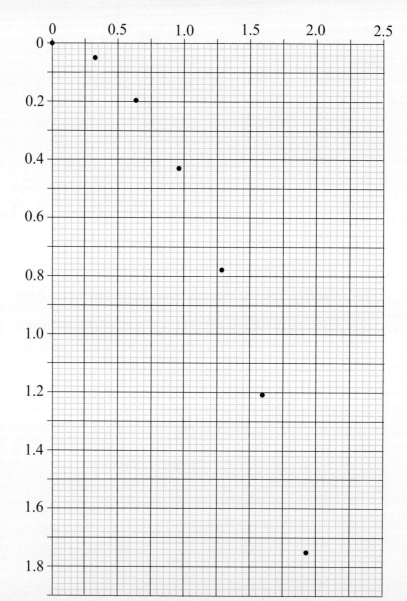

(a)   Use data from the diagram to determine the acceleration of free fall.   [3]

(b)   Determine the speed of the sphere 1.2 s after release.   [5]

(c)   On the grid, draw the path of the sphere assuming air resistance is not negligible.   [2]

B3. This question is in **two** parts. **Part 1** is about nuclear decay and ionization. **Part 2** is about radio waves.

Part 1 Nuclear decay and ionization

(a) A nucleus of radium-226 $\left(^{226}_{91}\text{Ra}\right)$ undergoes alpha particle decay to form a nucleus of radon (Rn).

Identify the proton number and nucleon number of the nucleus of Rn. *[2]*

Proton number:

Nucleon number:

(b) Immediately after the decay of a stationary radium nucleus, the alpha particle and the radon nucleus move off in opposite directions and at different speeds.

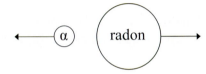

(i) Outline the reasons for these observations. *[3]*

(ii) Show that the ratio

$$\frac{\text{initial kinetic energy of alpha particle}}{\text{initial kinetic energy of radon atom}}$$

is about 56. *[3]*

(c) The initial kinetic energy of the alpha particle is 4.9 MeV. As the alpha particle passes through air, it loses all its kinetic energy by causing the ionization of 1.7×10^5 air molecules.

(i) State what is meant by ionization. *[1]*

(ii) Estimate, in joules, the average energy needed to ionize an air molecule. *[2]*

(d) Outline why a beta particle has a longer range in air than an alpha particle of the same energy. *[3]*

Part 2 Radio waves

A satellite is travelling directly away from a space shuttle with a speed of $7.9\,\text{km s}^{-1}$. The satellite is transmitting radio waves of frequency 150.000 000 MHz.

(a) (i) Calculate the frequency to which an astronaut on the shuttle will need to set a receiver in order to receive the signal from the satellite. *[2]*

(ii) The satellite now passes close to a second space shuttle. Describe the changes that will need to be made to the setting of the receiver as the satellite approaches and moves away from the second shuttle. *[3]*

(iii) Suggest why the formula used in the calculation in (a)(i) would **not** be suitable for radiation from a distant galaxy moving away from the Earth at 90% of the speed of light. *[1]*

(b)   A space shuttle orbits at a height of 300 km above the surface of the Earth. It carries two panels separated by a distance of 24 m. The panels reflect light of wavelength 500 nm towards an observer on the Earth's surface.

The observer views the panels with a telescope of aperture diameter 85 mm. The panels act as point sources of light for the observer.

    (i)   Describe what is meant by the Rayleigh criterion.   *[2]*

    (ii)   Determine whether the images of the panels formed by the telescope will be resolved.   *[3]*

B4.   This question is in **two** parts. **Part 1** is about an electrical heater. **Part 2** is about the hydrogen atom.

Part 1   Electrical heater

An electrical heater consists of two heating elements E_1 and E_2. The elements are connected in parallel. Each element has a switch and is connected to a supply of emf 240 V. The supply has negligible internal resistance.

Element E_1 is made from wire that has a cross-sectional area of $6.8 \times 10^{-8} \, \text{m}^2$. The resistivity of the wire at the operating temperature of the element is $1.1 \times 10^{-6} \, \Omega \, \text{m}$.

(a)   (i)   The total length of wire is 4.5 m. Show that the resistance of E_1 is 73 Ω.   *[1]*

    (ii)   Calculate the power output of E_1 with only this element connected to the supply.   *[2]*

    (iii)   Element E_2 is made of wire of the same cross-section and material as E_1. The length of wire used to make E_2 is 1.5 m. Determine the total power output when both E_1 and E_2 are connected to the supply.   *[3]*

    (iv)   With reference to the power output, explain why it would be inappropriate to connect the heating elements in series.   *[3]*

(b)   Each element in the electrical heater is wound as a coil as shown.

Each turn of the coil may be considered to act as a current-carrying long straight wire.

(i) On the diagram, draw the magnetic field around a current-carrying long straight wire. The arrow shows the direction of the current. [3]

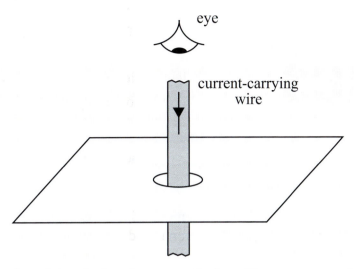

(ii) State **and** explain whether the turns of wire will attract or repel one another. [3]

Part 2 The hydrogen atom

(a) A parallel beam of visible light is shone through monatomic hydrogen gas.

The radiation emerging from the gas is analysed by comparing the incident and emerging intensities at various wavelengths. It is found that at a wavelength of 490 nm the intensity of the emergent beam is greatly reduced.

The diagram shows some of the electron energy states of the hydrogen atom where n is the quantum number of the energy level.

(i) Calculate the energy, in eV, of a photon of light of wavelength 490 nm. [2]

(ii) Use your answer in (a)(i) and the energy level diagram to explain the reduction in intensity of the emergent beam. [4]

(b) Outline how the Schrödinger model of the hydrogen atom leads to the idea of discrete electron energy states. [4]

Notes

Also from the IB store...

International Baccalaureate
Baccalauréat International
Bachillerato Internacional

Other subjects in the *IB Prepared* series

Group 3

Group 4

Group 5

Core requirements

Sign up to receive the IB store eNewsletter and hear about new subjects as they are added to this series

You may also be interested in... *IB Questionbank* series

Group 2

Group 3

Group 4

Group 5

Sign up to receive the IB store eNewsletter and hear about new subjects as they are added to this series

Get involved!

We welcome feedback on existing publications and any suggestions for new publications to complement IB programme materials:

- **Leave a review** on the relevant product page on the IB store
- **Send ideas and suggestions** for new resources to publishing.proposals@ibo.org

New publication alert/ eNewsletter sign-up

Visit the IB store to sign up for new publication alerts or to receive our quarterly eNewsletter.

Discounts

The more copies you buy the more you save —volume discounts now available on selected products

Stationery items and accessories

 Duo highlighter pen multipack

 Flower highlighter

 Laptop sleeve

 Baseball cap

For more items like these, go to the **Gift items** area of the IB store.

Downloads

Did you know that individual exam papers are available to buy on the IB store? To find exam papers for your subject area, go to the IB store at **http://store.ibo.org > Diploma Programme (DP) > Examinations, reports & markschemes.**

Many of our publications have sample chapters/pages available to download for free. See the product page on the IB store for details.

To make a purchase or for further information about IB products, prices and services, please visit the IB store at **http://store.ibo.org** or contact the IB store team at **sales@ibo.org**.